The Urban Book Series

The Urban Book Series is a resource for urban studies and geography research worldwide. It provides a unique and innovative resource for the latest developments in the field, nurturing a comprehensive and encompassing publication venue for urban studies, urban geography, planning and regional development.

The series publishes peer-reviewed volumes related to urbanization, sustainability, urban environments, sustainable urbanism, governance, globalization, urban and sustainable development, spatial and area studies, urban management, transport systems, urban infrastructure, urban dynamics, green cities and urban landscapes. It also invites research which documents urbanization processes and urban dynamics on a national, regional and local level, welcoming case studies, as well as comparative and applied research.

The series will appeal to urbanists, geographers, planners, engineers, architects, policy makers, and to all of those interested in a wide-ranging overview of contemporary urban studies and innovations in the field. It accepts monographs, edited volumes and textbooks.

Now Indexed by Scopus!

More information about this series at http://www.springer.com/series/14773

Zoé A. Hamstead · David M. Iwaniec ·
Timon McPhearson · Marta Berbés-Blázquez ·
Elizabeth M. Cook · Tischa A. Muñoz-Erickson
Editors

Resilient Urban Futures

Springer

Editors
Zoé A. Hamstead
Urban and Regional Planning
State University of New York
Buffalo, NY, USA

David M. Iwaniec
Urban Studies Institute
Georgia State University
Atlanta, GA, USA

Timon McPhearson
Urban Systems Lab
New School
New York, NY, USA

Marta Berbés-Blázquez
School for the Future of Innovation
in Society
Arizona State University
Tempe, AZ, USA

Elizabeth M. Cook
Department of Environmental Science
Barnard College
New York, NY, USA

Tischa A. Muñoz-Erickson
International Institute of Tropical Forestry
USDA Forest Service
Rio Piedras, PR, USA

ISSN 2365-757X ISSN 2365-7588 (electronic)
The Urban Book Series
ISBN 978-3-030-63130-7 ISBN 978-3-030-63131-4 (eBook)
https://doi.org/10.1007/978-3-030-63131-4

This Springer imprint is published by the registered company Springer Nature Switzerland AG
The registered company address is: Gewerbestrasse 11, 6330 Cham, Switzerland

Acknowledgments

The editors are grateful for the creativity, insights, and dedication of many people who contributed to this volume and the Urban Resilience to Extremes (UREx) Sustainability Research Network (SRN) project as a whole. We would like to thank all of the contributors to this volume, for sharing their conceptual and methodological frameworks that guide urban practice toward creating resilient and equitable urban futures. We are indebted to Lanika Sanders and Rakshanda Nagaraj, who carefully administered the process of compiling and copy-editing this work and without whom we would not have kept our deadlines. Twenty peer reviewers, including a student review panel, helped to improve the quality of the chapter manuscripts. The leadership and mentorship of Nancy Grimm and Chuck Redman have been invaluable to this project. In particular, their support of junior faculty, post-doctoral fellows and student trainees activated a substantial cohort of interdisciplinary researchers in cultivating an international community of practice around building more resilient and equitable futures. Far too many people to name co-created and participated in future scenarios workshops in Baltimore, Maryland, USA; Hermosillo, Mexico; Miami, Florida, USA; New York, New York, USA; Phoenix, Arizona, USA; Portland, Oregon, USA; Syracuse, New York, USA; San Juan, Puerto Rico, USA; and Valdivia, Chile. Over 220 practitioners and 180 researchers from 21 institutions in these cities dedicated their creativity, time, and expertise to building the resilience, future, and city-making knowledge base documented in this volume. We are grateful to the National Science Foundation which supported the UREx SRN via grant no. SES 1444755, GCR-1934933 (SETS Convergence) as well as the Long-Term Ecological Research Program Baltimore Ecosystem Study (BES) DEB-1637661, the Central Arizona-Phoenix (CEB-1637590) and the Chilean Science, Technology, Knowledge and Innovation Ministry of Chile (CONICYT-FONDECYT 3150290). Finally, we are grateful to our families for their support, kindness, not to mention patience with our travel schedules. They provide the grounding for this challenging, important work in building more equitable and resilient urban futures in the face of climate change.

Contents

Editors and Contributors

About the Editors

Zoé A. Hamstead is an Assistant Professor of Environmental Planning at the University at Buffalo and Director of the Community Resilience Lab. Her research builds on interdisciplinary approaches in urban planning, geography, and urban ecology to support climate equity and adaptive climate planning approaches. Dr. Hamstead teaches courses in environmental planning, environmental justice, economic concepts, geographic analysis of environmental problems, and engaged community energy studios. She holds a Ph.D. in Urban & Public Policy from The New School, a Master's in City & Regional Planning from the University of North Carolina at Chapel Hill, and a Bachelor of Liberal Arts from St. John's College.

David M. Iwaniec is an Assistant Professor of Urban Sustainability at the Urban Studies Institute, Andrew Young School of Policy Studies at Georgia State University. He is a sustainability scientist researching anticipatory and systems approaches to advance urban sustainability, resilience, and justice. His work focuses on the co-development of scenarios and transition pathways for positive futures of urban transformation.

Timon McPhearson is Director of the Urban Systems Lab and Associate Professor of Urban Ecology at The New School. He is a Research Fellow at The Cary Institute of Ecosystem Studies and Stockholm Resilience Centre and a member of the IPCC.

Marta Berbés-Blázquez is an Assistant Professor at the School for the Future of Innovation at Arizona State University. Her research considers the human dimensions of social-ecological transformations in rural and urban ecosystems with an emphasis on vulnerable populations. Her work is informed by resilience thinking and political ecology at a conceptual level, and it is practically oriented toward qualitative, participatory, and anticipatory research methods. Specific topics of expertise include power dynamics and access in ecosystem services, scenario planning, resource extraction, eco-health, climate change adaptation, and transformation.

Elizabeth M. Cook is an Assistant Professor at Barnard College in the Department of Environmental Science. She is an urban ecosystem ecologist and her research focuses on future urban sustainability and human–environment feedbacks in urban and nearby native ecosystems. She conducts research on sustainability and resilience planning through participatory scenario development with local stakeholders. Her work seeks to understand cities as social–ecological–technological systems with a comparative approach in Latin American and U.S. cities. Cook holds a Ph.D. from Arizona State University and a Bachelor of Arts from Wellesley College.

Tischa A. Muñoz-Erickson is a Research Social Scientist in the USDA Forest Service's International Institute of Tropical Forestry, in Río Piedras, Puerto Rico. She studies urban sustainability governance, including the policy networks, knowledge systems, anticipatory capacities, and strategies to advance sustainability, resilience, and equity. She is also actively involved in transdisciplinary platforms to facilitate the co-production of futures and transition pathways in the US and Latin American cities. Muñoz-Erickson leads the International Urban Field Station and the San Juan ULTRA and is also Co-PI of the National Science Foundation Urban Resilience to Extreme Events Sustainability Research Network (UREx SRN) and the SETS Convergence project.

Contributors

Marta Berbés-Blázquez School for the Future of Innovation in Society, Arizona State University, Tempe, AZ, USA

Rocio Carrero Urban Systems Lab, The New School, New York, NY, USA

Elizabeth M. Cook Department of Environmental Science, Barnard College, New York City, NY, USA

Mathieu Feagan School for the Future of Innovation in Society, Arizona State University, Tempe, AZ, USA

Nancy Grimm School of Life Sciences, Arizona State University, Arizona, USA; Julie Ann Wrigley Global Institute of Sustainability, Arizona State University, Tempe, AZ, USA

Nancy B. Grimm School of Life Sciences, Arizona State University, Tempe, AZ, USA

Zoé A. Hamstead Department of Urban and Regional Planning, School of Architecture and Planning, University at Buffalo, State University of New York, Buffalo, New York, NY, USA

Robert Hobbins School of Sustainability, Arizona State University, Tempe, AZ, USA

David M. Iwaniec Andrew Young School of Policy Studies, Urban Studies Institute, Georgia State University, Atlanta, GA, USA

Yeowon Kim Julia Ann Wrigley Global Institute of Sustainability, Arizona State University, Tempe, AZ, USA

Lelani Mannetti Urban Studies Institute, Andrew Young School of Policy Studies, Georgia State University, Georgia, USA

Lelani M. Mannetti Andrew Young School of Policy Studies, Urban Studies Institute, Georgia State University, Atlanta, GA, USA

Samuel Markolf Julia Ann Wrigley Global Institute of Sustainability, Arizona State University, Tempe, AZ, USA

Timon McPhearson The New School, Urban Systems Lab, New York, NY, USA;
Cary Institute of Ecosystem Studies, Millbrook, NY, USA;
Stockholm Resilience Center, Stockholm University, Stockholm, Sweden

Clark Miller School for the Future of Innovation in Society, Arizona State University, Tempe, AZ, USA

Thaddeus R. Miller School of Public Policy, University of Massachusetts Amherst, Amherst, USA

Tischa A. Muñoz-Erickson USDA Forest Service, International Institute of Tropical Forestry, Río Piedras, PR, USA

A. Mustafa Urban Systems Lab, The New School, New York, NY, USA

L. Ortiz Urban Systems Lab, The New School, New York, NY, USA

Jaskirat Randhawa The New School, Urban Systems Lab, New York, NY, USA

B. Rosenzweig Environmental Science, Natural Sciences and Mathematics, Sarah Lawrence College, Bronxville, NY, USA

Jason Sauer School of Life Sciences, Arizona State University, Tempe, AZ, USA

Daniel Sauter The New School, Urban Systems Lab, New York, NY, USA

Kaethe Selkirk School for the Future of Innovation in Society, Arizona State University, Tempe, AZ, USA

Claudia Tomateo The New School, Urban Systems Lab, New York, NY, USA

Vivian Hobbins Instituto Tecnológico de Sonora, Sonora, Mexico, USA

Darin Wahl Lund University Centre for Sustainability Studies, Lund, Sweden

Chapter 1
A Framework for Resilient Urban Futures

David M. Iwaniec, Nancy B. Grimm, Timon McPhearson, Marta Berbés-Blázquez, Elizabeth M. Cook, and Tischa A. Muñoz-Erickson

Abstract Resilient urban futures provides a social–ecological–technological systems (SETS) perspective on promoting and understanding resilience. This chapter introduces the concepts, research, and practice of urban resilience from the Urban Resilience to Extremes Sustainability Research Network (UREx SRN). It describes conceptual and methodological approaches to address how cities experience extreme weather events, adapt to climate resilience challenges, and can transform toward sustainable and equitable futures.

Keywords Urban futures · Co-production · Resilience · Scenario visions · Positive futures

1.1 Introduction

If one were to imagine that each time a disaster or stress strikes people on the earth, a strong beacon would illuminate—like an alert board but extended across the globe—then the cities of the world would frequently light up. Cities would give such

D. M. Iwaniec (✉)
Andrew Young School of Policy Studies, Urban Studies Institute, Georgia State University, Atlanta, GA, USA
e-mail: diwaniec@gsu.edu

N. B. Grimm
School of Life Sciences, Arizona State University, Tempe, AZ, USA

T. McPhearson
The New School, Urban Systems Lab, New York, NY, USA

M. Berbés-Blázquez
School for the Future of Innovation in Society, Arizona State University, Tempe, AZ, USA

E. M. Cook
Department of Environmental Science, Barnard College, New York, NY, USA

T. A. Muñoz-Erickson
USDA Forest Service, International Institute of Tropical Forestry, Rio Piedras, PR, USA

Z. A. Hamstead et al. (eds.), *Resilient Urban Futures*, The Urban Book Series,
https://doi.org/10.1007/978-3-030-63131-4_1

a strong signal because they are often the places most vulnerable to disasters. Cities concentrate people and infrastructure. They are often located along coasts or rivers and are thus exposed to floods or tropical storms, and are susceptible to drought, fire, and heat. This exposure combined with poor infrastructural and institutional adaptation can mean the difference between a hazard and a disaster. Furthermore, the frequency, intensity, and impact of such events are increasing as human-caused emissions of heat-trapping gases like carbon dioxide and methane continue to rise, spurring changes in the earth's climate system. The coupling of a major demographic transition to urban living with climate change, especially the increase in frequency and intensity of extreme events, presents an increasingly urgent challenge to urban society and decision makers.

In the context of climate risk for cities, interest in the concept of resilience and its application to urban systems has exploded. Resilience has many definitions (Meerow et al. 2016), but for our purposes we open with one that comes from ecology. Resilience is the capacity of a system to maintain its basic structure, function, and identity while undergoing change in the face of shocks and stresses (Walker et al. 2004; Folke et al. 2010). In this definition, resilience is seen as a property of a dynamic system (Elmqvist et al. 2019). Other definitions point to resilience as an outcome or process (Moser et al. 2019). The processes and outcomes that define resilience in an urban social–ecological–technological system (SETS) have to do with deliberate management of that system to build or promote its resilience through adaptation and transformation (Pelling 2010; Biggs et al. 2012).

1.2 An Approach to Urban Resilience Research-Practice

Resilience has gained status as a guiding concept in urban planning and management. In this book, a group of researchers from the Urban Resilience to Extremes Sustainability Research Network (UREx SRN) considers the past, present, and future challenges for cities to build resilience. The UREx SRN is a network of collaborating interdisciplinary researchers and practitioners from diverse world cities working together to promote, design, and implement urban infrastructure that is resilient in the face of future extreme events, provides ecosystem services, improves social wellbeing, and exploits new technologies in ways that benefit all segments of urban populations.

The network consists of over 180 researchers from 21 institutions working together with over 220 practitioners in nine cities of the United States and Latin America. Researchers, including faculty members, staff scientists, post-doctoral fellows, and graduate fellows and associates, represent disciplines within engineering, life, physical, and social sciences. Practitioners include governmental officials (city, county, and state level), community organization leaders, non-governmental organizations, business leaders, and others who are involved to varying extents in decision-making. The cities include Miami, FL; Baltimore, MD; New York, NY; Syracuse, NY; Portland, OR; Phoenix, AZ; and, in Latin America and the Caribbean, Hermosillo, Mexico; Valdivia, Chile; and San Juan, Puerto Rico.

The UREx SRN approach to resilient urban futures has two key elements: SETS and network interactions. The UREx SRN has developed SETS into a comprehensive epistemological framework for understanding and promoting resilience. In this framework, the social, ecological, and technological components of a system all are considered, and resilient strategies integrate all three domains and their interactions. For example, a coastal city like Miami that is experiencing sea-level rise and is exposed to hurricanes and associated storm surges and wind considers coastal fortification, elevating urban surfaces, and pumping out sea water (technological), but may also opt to restore coastal mangrove ecosystems (ecological). These strategies may be adopted by some municipalities but not others (governance; social), perhaps owing to differences in municipal revenue (economics; social). People living in the Miami area are differentially exposed to nuisance tidal flooding, in part as a function of the geographical/topographic setting (ecological) that may be associated with differences in housing prices (social) or the degree to which protective infrastructure (technological) is available. Vulnerability to more severe coastal flooding also varies, with lower-income, marginalized communities—often predominantly Black or Latinx—most at risk (social). These examples illustrate the high degree of spatial and social heterogeneity in who and what is most vulnerable, as well as variability in the number and types of interventions and their SETS interactions that are meant to build resilience against coastal inundation.

Most approaches to improving resilience are siloed, with efforts focused on one or, rarely, two domains. Yet extreme events often cause cascading impacts (Rocha et al. 2018). For example, flooding can simultaneously cause power and transportation disruptions, damage ecosystems, impact human health, and damage homes and critical infrastructure. Recent hurricanes demonstrated failures or inadequacies not only in built infrastructure but also in resources, institutions, and information systems—components of the urban SETS—to prepare for and respond to events of this magnitude (Eakin et al. 2018; Markolf et al. 2018). Solutions that address only one system domain are unlikely to prove resilient in the future. Because impacts occur among interdependent human, climate-biophysical, engineered systems, we suggest a fundamental rethinking, reanalysis, and remaking of cities as social–ecological–technological systems (SETS).

By adopting the SETS framework, the UREx SRN places priority on integrating equity considerations into all projects. Consideration of equity asks us to examine the legacies and continued impact of discrimination, such as redlining in urban planning or discriminatory banking or housing policies and practices, that have produced differential vulnerability today. It also demands that voices of those most affected by potential actions to build resilience are privileged, and that actions proposed or taken do not exacerbate unfair practices.

UREx SRN researchers have advanced the concept of safe-to-fail infrastructure (Ahern 2011; Kim et al. 2017, 2019), which may incorporate natural elements but principally is a new way of thinking about infrastructure under uncertain and changing probabilities of extreme events. In this framing, instead of being designed to withstand events of a certain magnitude, infrastructure systems are conceived as flexible, multifunctional, and able to adapt or transform through the interactions of

S, E, and T. In fact, these interactions are key to thinking about the future and ways to manage for or enhance resilience to increasing frequency and magnitude of extreme events under the SETS framework. For example, green gentrification may accompany the establishment of nature-based strategies (an example might be green swales and rain gardens for stormwater management), a process by which historically disenfranchised populations are pushed out by neighborhood improvements that increase cost of living, or are excluded from green infrastructure planning processes (Kabisch and Haase 2014; Locke and Grove 2016; Hobbie and Grimm 2020). Thinking in terms of SETS requires that future makers consider the social implications of ecological or technological interventions, as well as how social institutions at various scales influence the success of such interventions. Future makers include the wide variety of actors and the governance venues in which residents engage in creating change. These policy actors work with historical narratives, current conditions, and imaginative tools to create visions for the future. A primary research question that the UREx SRN project asks of urban systems science is: how do SETS domains interact to generate vulnerability or resilience, and how can urban SETS dynamics be guided along more resilient, equitable, and sustainable trajectories?

The UREx SRN has developed a strong network of researchers while promoting interaction between researchers and practitioners, and among practitioners from different cities. A focus on positive futures is achieved through scenarios co-produced in each city through participatory workshops engaging diverse practitioners. This co-production of knowledge and action enables incorporation of diverse sectoral, cultural, and disciplinary viewpoints into plausible and desirable future visions for each city. Because UREx SRN has included Latin American cities, shared learning across all UREx SRN cities provides increased capacity and diversity of perspectives on resilience (e.g., information to cities that are experiencing a rise in Hispanic populations), and network-level opportunities for collaboration and collective action (e.g., sharing best practices across international networks).

1.3 Linking the Past, Present, and Future

Throughout this book, the authors discuss conceptual and methodological approaches to address how cities experience extreme weather events, adapt to resilience challenges, and can transform toward sustainable and equitable futures while contributing to urban systems science. This book seeks to advance an urban system research-practice that brings together diverse knowledge, skills, tools, perspectives, and ideas. The objective of this transdisciplinary approach is to build actionable, anticipatory knowledge for decision-making. Each city in the UREx SRN has a researcher–practitioner team that works together to plan research and participatory scenario workshops to co-produce positive visions of those cities' futures. These scenario visions are meant to provide information on how different goals, strategies, and targets can work in synergy or may require trade-offs in decision-making. The first section, Chaps. 2–5, focuses on understanding urban S, E, and T vulnerabilities and resilience.

The second section, Chaps. 6–12, then applies these insights to an innovative framework that guides city stakeholders toward integrating the three domains along more sustainable, equitable, and resilient trajectories.

In Chap. 2, Hamstead argues that we cannot transform cities for future generations unless we unpack ways in which economic and political institutions have created the climate crisis and inscribed climate inequity into the urban built environment. She describes these trends broadly, and uses the example of Puerto Rico to trace ways in which colonialism, land use change, and scientific and political narratives all work together to reinforce societal inequity.

In Chap. 3, Kim et al. describe a methodology to characterize how cities define and prioritize climate adaptation strategies in governance. The SETS framework is used as a lens to explore the dynamics and interrelationships of the goals, solutions, and targets put forth in formal planning documents. This chapter provides a codebook for doing content analysis of municipal planning documents to explore the diverse SETS strategies cities are employing to address climate resilience, specifically related to extreme weather events such as heat, drought, and flooding. The proposed SETS governance analysis helps stakeholders understand how urban planning addresses current and future climate vulnerabilities and explores the various adaptation options that cities have prioritized for the community.

In Chap. 4, Hamstead and Sauer look at how spatial patterns of vulnerability to extreme events are manifestations of structural injustice that leave their mark on the built environment, and the ways in which socio-spatial segregation patterns are aligned with patterns of exposure to extreme events. Spatial vulnerability assessments can be powerful tools for prioritizing where and how cities should make investments for mitigating the impacts of extreme events, and can provide an entry point for asking more fundamental questions about the processes that produce patterns of climate inequity. Using commensurate indicators of exposure, sensitivity, and adaptive capacity, flood and heat vulnerability mapping is used to convey distributional patterns (e.g., mapping urban landscapes and extreme event injustice) that enable communities to identify hazardous biophysical conditions and residents who are most at risk of exposure to those conditions.

In Chap. 5, Hobbins et al. argue that, for a knowledge system to produce quality knowledge for decision-making, it requires more than the best scientific data and the most sophisticated technology; the distribution of power and authority also dramatically influences the quality, legitimacy, and accuracy of the knowledge claims produced by a knowledge system. This chapter demonstrates the value of knowledge systems analysis as a method to stress-test and identify weaknesses of a knowledge system that warrant attention. Knowledge systems analysis can inform potential solutions to upgrade or redesign a knowledge system in support of building resilient cities. This is illustrated through an analysis of the United States Federal Emergency Management Agency (FEMA) Flood Insurance Rate Map knowledge system and sheds light on the underlying social and political dynamics involved in how we know, review and validate, communicate, and use flood risk knowledge.

In Chap. 6, Iwaniec et al. present the UREx SRN framework for scenario development of positive urban futures. Three distinct scenario approaches are used to explore potential outcomes of existing planning goals (strategic scenarios), to craft visions that address pressing resilience challenges (adaptive scenarios), and to articulate visions of radical departures from the status quo in the pursuit of resilience, sustainability, and equity (transformative scenarios). The scenarios are developed in participatory workshop settings designed to build anticipatory capacity, or very long-term forward thinking. A series of creative and analytical processes are used to engage the community in imagining, articulating, and scrutinizing visions and pathways of positive futures. The approach offers an alternative and complement to traditional forecasting techniques by applying inspirational stories to resilience research and practice.

In Chap. 7, Cook et al. describe the process of co-production in which UREx SRN participatory scenario development takes place. Key characteristics of meaningful co-production are highlighted to draw attention to the benefits of engaging in collaborative knowledge production and co-generation of resilient urban futures with diverse perspectives. The chapter focuses on centering a collective commitment, enhancing credibility, legitimacy, and accountability, and empowering diverse perspectives through an iterative, flexible process. It also reflects on challenges that must be considered in an engagement, co-production process and the lessons learned from the UREx SRN project.

In Chap. 8, Berbés-Blázquez et al. demonstrate an approach to assess and compare co-produced scenario visions, which consists of a multi-criteria assessment used to explore the resilience, equity, and sustainability dimensions reflected in scenario visions in a qualitative manner (RESQ). This is illustrated by applying the assessment to compare heat and drought visions from Hermosillo and Phoenix. However, the approach described in this chapter is not intended merely as a tool to assess the strengths and weaknesses of a given vision. Qualitative evaluation of a scenario vision (and comparisons among scenario visions) offers an opportunity for reflection and dialog on underlying values and aspirations. The RESQ assessment is also presented as an initial step toward developing futures-oriented indicators of resilience, equity, and sustainability for cities.

In Chap. 9, Ortiz et al. introduce land use modeling techniques to produce and evaluate spatially explicit urban futures via the UREx SRN scenario co-production process. Weather hazards, projected to become more frequent and intense, pose critical threats to cities and the people in them. The complexity and scale of these threats will require adaptation strategies that meet their scale. This chapter presents data-driven techniques to estimate impacts of land use change on heat and flood risks in cities that combine statistical and process-based approaches. These approaches quantify heat and flood hazard as a function of the urban landscape that responds to a large-scale climate signal. Co-produced future land use scenarios can then be evaluated on their impacts on heat and flood hazard following the techniques provided in the chapter.

In Chap. 10, Sauter et al. present a data visualization approach as an interactive web application to visualize urban SETS. The platform was conceived as a tool to produce anticipatory knowledge, bridging the gap between quantitative social, ecological and infrastructure data, and the rich and layered qualitative insights compiled from local stakeholder scenario visioning workshops. The objective is to support the exploration and understanding of complex geospatial relationships for use in research and decision support, and to do so in a simple and organized way while avoiding visual outputs that cognitively overwhelm the viewer.

In Chap. 11, Muñoz-Erickson et al. make the case for anticipation as a critical component of building resilience and the need to embed anticipatory thinking in urban planning practices and knowledge systems. This chapter introduces "anticipatory resilience" as a futures-oriented knowledge system that intentionally explores alternative, desirable future states and suggests a portfolio of tools suitable for building long-term foresight capacity in urban planning. Examples of knowledge systems interventions are presented to explore the trade-offs, constraints, possibilities, and desires of diverse future scenarios co-generated in settings with people from different perspectives, knowledge, and expectations.

The book concludes with a vision for advancing the science and practice of co-producing positive, urban futures. This final chapter discusses the importance of systems thinking, the need to advance development of an urban systems science but also an urban systems practice, and why positive visioning is key to counter the dystopian narratives and scenarios that dominate discourses of our shared urban future. The need for inclusive and diverse engagement and the recognition of the privilege of both those who are able to do urban futures work and also those that tend to be included in co-production is provided as a key learning from the UREx SRN work as well as a call to action for more just and inclusive processes in envisioning and planning more resilient urban futures.

References

Ahern J (2011) From fail-safe to safe-to-fail: sustainability and resilience in the new urban world. Landscape Urban Plann 100(4):341–343. https://doi.org/10.1016/j.landurbplan.2011.02.021

Biggs R, Schlüter M, Biggs D et al (2012) Toward principles for enhancing the resilience of ecosystem services. Annu Rev Environ Resour 37:421–448. https://doi.org/10.1146/annurev-environ-051211-123836

Eakin H, Muñoz-Erickson TA, Lemos MC (2018) Critical lines of action for vulnerability and resilience research and practice: Lessons from the 2017 hurricane season. J Extreme Events 05(02n03):1850015. https://doi.org/10.1142/S234573761850015X

Elmqvist T, Andersson E, Frantzeskaki N et al (2019) Sustainability and resilience for urban transformations. Nat Sustain 2:267–273. https://doi.org/10.1038/s41893-019-0250-1

Folke C, Carpenter SR, Walker B et al (2010) Resilience thinking: Integrating resilience, adaptability and transformability. Ecol Soc 15(4):20

Hobbie SE, Grimm NB (2020) Nature-based approaches to managing climate change impacts in cities. Phil Trans R Soc B 375:20190124. https://doi.org/10.1098/rstb.2019.0124

Kabisch N, Haase D (2014) Green justice or just green? Provision of urban green spaces in Berlin, Germany. Landscape Urban Plann 122:129–139. https://doi.org/10.1016/j.landurbplan.2013.11.016

Kim Y, Chester MV, Eisenberg DA et al (2019) The infrastructure trolley problem: positioning safe-to-fail infrastructure for climate change adaptation. Earth's Future 7(7):704–717. https://doi.org/10.1029/2019EF001208

Kim Y, Eisenberg DA, Bondank EN et al (2017) Fail-safe and safe-to-fail adaptation: decision-making for urban flooding under climate change. Clim Change 145(3–4):397–412. https://doi.org/10.1007/s10584-017-2090-1
Locke DH, Grove JM (2016) Doing the hard work where it's easiest? Examining the relationships between urban greening programs and social and ecological characteristics. Appl Spatial Anal Policy 9:77–96. https://doi.org/10.1007/s12061-014-9131-1
Markolf SA, Chester MV, Eisenberg DA et al (2018) Interdependent infrastructure as linked social, ecological, and technological systems (SETSs) to address lock in and enhance resilience. Earth's Future 6(12). https://doi.org/10.1029/2018EF000926
Meerow S, Newell JP, Stults M (2016) Defining urban resilience: a review. Landscape Urban Plann 147:38–49
Moser S, Meerow S, Arnott J (2019) The turbulent world of resilience: Interpretations and themes for transdisciplinary dialogue. Clim Change 153:21–40. https://doi.org/10.1007/s10584-018-2358-0
Pelling M (2010) Adaptation to climate change: from resilience to transformation. Routledge, London and New York
Rocha JC, Peterson G, Bodin Ö et al (2018) Cascading regime shifts within and across scales. Science 362(6421):1379–1383. https://doi.org/10.1126/science.aat/7850
Walker B, Holling CS, Carpenter SR et al (2004) Resilience, adaptability and transformability in social–ecological systems. Ecol Soc 9(2):5. https://doi.org/10.5751/ES-00650-090205

David M. Iwaniec is an Assistant Professor of Urban Sustainability at the Urban Studies Institute, Andrew Young School of Policy Studies at Georgia State University. He is a sustainability scientist researching anticipatory and systems approaches to advance urban sustainability, resilience, and justice. His work focuses on the co-development of scenarios and transition pathways for positive futures of urban transformation

Nancy B. Grimm studies the interactions between climate variation and change, human activities, and ecosystems. Her interdisciplinary urban and stream ecosystem research focuses on disturbance, resilience, and biogeochemical processes. Dr. Grimm is Regents Professor at Arizona State University and currently co-directs the Urban Resilience to Extremes Sustainability Research Network (UREx SRN). She was president (2005-06) and is a fellow of the Ecological Society of America, as well as a fellow of the American Geophysical Union (AGU), the American Association for the Advancement of Science (AAAS), and the Society for Freshwater Science. She is an elected member (2019) of the US National Academy of Sciences.

Timon McPhearson is the Director of the Urban Systems Lab and Associate Professor of Urban Ecology at The New School. He is a Research Fellow at the Cary Institute of Ecosystem Studies and Stockholm Resilience Centre and a member of the IPCC

Marta Berbés-Blázquez is an Assistant Professor at the School for the Future of Innovation at Arizona State University. Her research considers the human dimensions of social–ecological transformations in rural and urban ecosystems with an emphasis on vulnerable populations. Her work is informed by resilience thinking and political ecology at a conceptual level, and it is practically oriented toward qualitative, participatory, and anticipatory research methods. Specific topics of expertise include power dynamics and access in ecosystem services, scenario planning, resource extraction, eco-health, climate change adaptation, and transformation.

Elizabeth M. Cook is an Assistant Professor at Barnard College in the Department of Environmental Science. She is an urban ecosystem ecologist and her research focuses on future urban sustainability and human–environment feedback in urban and nearby native ecosystems. She conducts research on sustainability and resilience planning through participatory scenario development with local stakeholders. Her work seeks to understand cities as social–ecological–technological systems with a comparative approach in Latin American and U.S. cities. Cook holds a Ph.D. from Arizona State University and a Bachelor of Arts from Wellesley College

Tischa A. Muñoz-Erickson is a Research Social Scientist in the USDA Forest Service International Institute of Tropical Forestry, in Río Piedras, Puerto Rico. She studies urban sustainability governance, including the policy networks, knowledge systems, anticipatory capacities, and strategies to advance sustainability, resilience, and equity. She is also actively involved in transdisciplinary platforms to facilitate the co-production of futures and transition pathways in the U.S. and Latin American cities. Muñoz-Erickson leads the International Urban Field Station and the San Juan ULTRA and is also Co-PI of the National Science Foundation Urban Resilience to Extreme Events Sustainability Research Network (UREx SRN) and the SETS Convergence project.

Chapter 2
How We Got Here: Producing Climate Inequity and Vulnerability to Urban Weather Extremes

Zoé A. Hamstead

Abstract This collected volume is intentionally future-oriented; it is authored by a team of interdisciplinary scientists and practitioners who collaborate to translate research findings into networked adaptive practices that we hope will protect urban communities against the impacts of extreme weather. While future-oriented, we cannot protect future generations against urban weather extremes without understanding the historical processes through which these existential and ethical crises came about. This chapter describes how economic and political institutions produced the climate crisis in ways that also constitute a humanitarian crisis, inscribing climate inequity into the urban built environment and institutions. It offers reflections on ways in which this history must be wrestled with in the context of equitable and resilient urban futures.

Keywords Urbanization · Extreme weather · Climate inequity · Procedural injustice · Distributional injustice

Using the example of Hurricane María in Puerto Rico and more general trends, this chapter illustrates how particular processes of economic change, ecological change, national and international policy, and power shape experiences with weather extremes. Understanding how crises like Hurricane María are produced involves more than scientific descriptions of global warming and extreme weather in relation to historical averages. It involves more than accounting for outcomes in terms of recovery cost, mortality rates, and climate migration predictions, or even differentiating these across geographies and social groups. It requires establishing how institutions at multiple scales and in multiple linked (social–ecological–technological domains [see Chap. 3]) inscribe inequality. Historically, climate change has its roots in economic change—the industrialization and carbonization of our economic processes, with concomitant development of cities. Owing to ways in which their biophysical landscape was designed, it is cities that are now particularly susceptible

Z. A. Hamstead (✉)
University at Buffalo, State University of New York, Buffalo, New York, NY, USA
e-mail: zoehamst@buffalo.edu

Z. A. Hamstead et al. (eds.), *Resilient Urban Futures*, The Urban Book Series,
https://doi.org/10.1007/978-3-030-63131-4_2

to heat islands, sea level rise, and extreme weather. But the question of which cities are able to harness resources necessary for recovery and which social groups within cities receive those resources is a matter of politics, power, and exclusionary practices. Examining how the social separateness that forms unequal climate burden is embedded in planning, policy, scientific, and economic practices is crucial for transforming these institutions in ways that protect vulnerable communities from extreme weather.

Vulnerabilities to extreme weather events are driven by interactions between exposure to the hazard (e.g., coastal location of infrastructure) and capacities for coping (e.g., ability to mobilize recovery resources). (For more information about vulnerability to weather extremes, see Chap. 4.) As global trade and other carbon-intensive activities increase exposure to extreme events globally, geographic differences make those exposures uneven, while economic, political, and other forms of social separateness drive differential abilities to cope with the impacts. Vulnerability is often described in terms of distributional injustice—the spatial, temporal, and demographic patterns of environmental burden and benefit. However, less attention is paid to conditions of recognition, respect, and fair and inclusive processes which are preconditions for distributional injustices, or ways in which distributional injustices exacerbate procedural injustices in turn. Procedural injustices committed by the state and other institutions are related to exclusionary practices that assign respect and recognition, (dis)enabling participation in political processes (Schlosberg 2007). Procedural justice can, therefore, be seen as a tool to achieve more equitable material distributions. All these forms of injustice play a role in vulnerability to weather extremes. Here, I use environmental justice theory and historical context to explain climate injustice. First, I describe broad trends in weather extremes and ways in which they are influenced by urbanization. I then describe various roles of power and politics in producing an unequal climate burden.

2.1 Breaking Climatological Records

In early September of 2017, category 5 Hurricane Irma brought wind gusts of 74 miles per hour (mph) [119 km per hour (kph)], storm surge flooding along the north coast, and heavy rainfall in the central and eastern portions of the island of Puerto Rico. Three weeks later, category 4 Hurricane María made landfall, bringing maximum sustained winds of 175 mph (282 kph). Along the northern, eastern, and southeastern coastlines, maximum storm inundation reached 10 feet (ft [3 m]), with heavy rainfall peaking at 37.9 inches (96.3 cm) (Fig. 2.1). Riverine flooding displaced communities along the Guajataca River, and over 2,000 people had to be rescued from rooftops in the municipality of Toa Baja (Ferré-Sadurní 2017) on the northern part of the island. Flooding also caused heavy metal and bacterial contamination of drinking water. Most of Puerto Rico's toxic waste sites—some of which contain coal ash comprised of heavy metals like lead and arsenic—are located along the southeastern coastline, which was heavily inundated (Funes 2017). Since more than a third of the sewerage treatment plants did not function after the hurricane and water service restoration

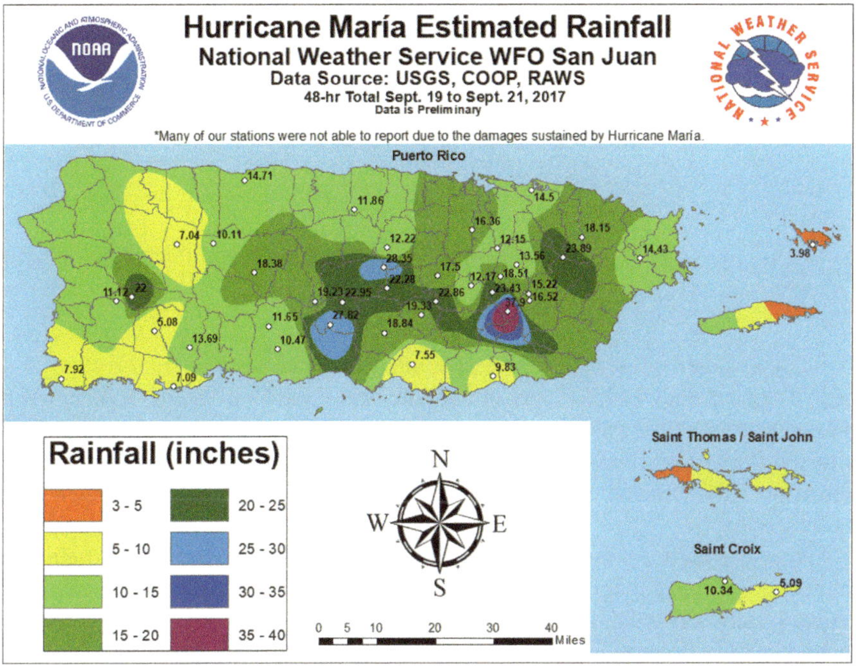

Fig. 2.1 Estimated rainfall totals from Hurricane María in Puerto Rico and the U.S. Virgin Islands with totals from reporting stations. Image credit: NOAA

was slow in many parts of the island, some people depended on waterways mixed with raw sewerage for drinking and bathing.

Damage to critical infrastructure cut off services for months as the death toll mounted and a mass migration to the United States mainland ensued. One month after the hurricane, more than 92% of Puerto Rico's roads were closed due to massive landslides and downed trees, leaving people struggling to get medical supplies, food, and water. In the storm's aftermath, Puerto Ricans experienced the largest blackout in U.S. history; 1.5 million residents were without power, which was not fully restored for almost a year. Overall, the Federal Emergency Management Commission (2018) estimates that María caused US $90 billion in damages, making it the third costliest storm in U.S. history (Table 2.1). In the year following the storm, an estimated 160,000 Puerto Ricans had relocated to the mainland (Hinojosa and Meléndez 2018).

U.S. Caribbean islands like Puerto Rico are sensitive to large-scale weather patterns such as the El Nino-Southern Oscillation and the Atlantic Multidecadal Oscillation, which can lead to weather and climate extremes (Gould et al. 2018). Since 1950, annual average temperatures in Puerto Rico have increased by about 1.5°F, or approximately 0.8 °C. Under a high emissions scenario, temperatures are projected to increase by as much as 9°F (5 °C) and sea level rise is expected to increase 9–11 ft (2.7–3.4 m) by 2100. In the Caribbean, these climatic changes will likely lead to water shortages, as some locations experience longer dry seasons and

Table 2.1 Damage estimates for costliest storms in U.S. history

Hurricane	Year	Cost (billion)
Katrina	2005	$161
Harvey	2017	$125
María	2017	$90
Sandy	2012	$71
Irma	2017	$50

Source Federal Emergency Management Commission (FEMA 2018)

shorter, but wetter wet seasons. Ocean warming and acidification will threaten coral reefs. The frequency and intensity of extreme heat events will increase. In addition to large-scale weather patterns, the coastal orientation of human settlements, with critical infrastructure, ecosystems, and economic activity in Caribbean islands make them especially vulnerable to destruction and human health impacts associated with severe storms.

Across the globe, communities are experiencing unstable weather and climate conditions unfavorable for the wellbeing of humans, and the ecological and technological systems on which they rely. Scientists describe extreme weather and climate phenomena in reference to historic averages; they exist on the outer boundaries of meteorological distributions, or the ranges of precipitation, temperature and other weather phenomena that have been observed in the past. Since these distributions are based on a location's recorded weather history, what is typical in one place may be extreme in another. The Intergovernmental Panel on Climate Change (IPCC) defines *extremes of atmospheric weather and climate variables* as:

> The occurrence of a value of a weather or climate variable above (or below) a threshold value near the upper (or lower) ends of the range of observed values of the variable (IPCC 2012, p. 557).

As climate change shifts the mean distributions of weather and climate, human adaptation is not keeping pace, causing injury to countless communities across the globe. Extreme weather events—including heat waves, droughts, heavy precipitation events, hurricanes, tropical storms, wildfires, landslides, and mudslides—have all caused irreversible damage in the last decade. The 14th edition of the Climate Risk Index analysis produced by Germanwatch reports that between 1998 and 2017, more than 526,000 deaths and U.S. $3.47 trillion (in purchasing power parities[1]) of damages were directly caused by over 11,500 extreme weather events (Eckstein et al. 2019). Globally, weather-related disasters have more than tripled since the 1960s (World Health Organization 2018). In some parts of the world, it is estimated that human-induced climate change has more than doubled the probability of heat waves and it will not be possible to reverse a large proportion of climate change within

[1]Purchasing power parity is based on different countries' relative prices levels for the same basket of goods, and provides a way to translate countries' currencies into U.S. Dollars (Dornbusch 1985).

the next 100 to 1,000 years (IPCC 2014). Even if humans were to stop emitting greenhouse gases, warming will continue to undermine human survival.

Climate-induced extreme events exacerbate existing vulnerabilities in ecosystems, infrastructure, economic systems, and human health. The agricultural sector is particularly vulnerable to climate change-induced drought in lower latitudes (IPCC 2007), threatening national economies and pushing farmers off their land. Extreme weather causes disruptions to infrastructure like water systems, sewer systems, roads, and power plants, already aging and in need of repair (USGCRP 2016). The United Nations High Commissioner for Refugees estimates that 22.5 million people were displaced by climate or weather-related events between 2008 and 2015, equivalent to 62,000 people each day (Yonetani et al. 2015). Weather extremes are occurring with levels of frequency, duration, and intensity for which ecosystems, human health, institutions, and technological infrastructure are ill-adapted.

According to the IPCC (Seneviratne et al. 2012), we can expect substantial warming in temperature extremes throughout the twenty-first century, especially over land areas. Particularly in high latitudes and tropical regions, the proportion of total rainfall from heavy precipitation events will increase and, in some regions, those increases will occur despite total decreases in precipitation. These changes will impact the natural environment in ways that drive droughts, floods, extreme sea level, waves, coastal impacts, cryosphere-related impacts, landslides, and sand and dust storms.

With high concentrations of people and engineered materials in locations which are often coastal, cities are particularly susceptible to extreme weather events and their impacts. Urbanization can describe many human settlement processes and patterns—from urban densification to suburban and exurban sprawling land use patterns. These forms involve altering ecosystems and replacing them with engineered infrastructure, which exacerbates extreme atmospheric events (Hamstead and Coseo 2020). During the mid to late nineteenth century, Europe and the United States underwent a period of industrial urbanization. This economic transformation marked a shift to carbon-intensive activities and land use patterns which not only led to global warming but also created unique vulnerabilities in the sites of that carbon-intensive activity.

2.2 Urbanization and Extreme Weather Events

Long before the industrial period, colonization and subsequent processes of economic change brought major transformations to the ecology and human settlement patterns of Puerto Rico. Before Europeans began to seize control of the island in the late fifteenth century, the Taíno people grew crops such as corn, yucca, yams, and cotton. Mangroves—coastal shrubs and trees that are crucial for shoreline protection and soil stabilization—were prevalent and widely distributed along the coast of the island. During this colonial period, sugar was becoming one of the most valuable European trade commodities. Thus, in Puerto Rico and elsewhere, land was cleared to make way for sugarcane and slaves were brought to fulfill labor demands. As sugarcane, tobacco, citrus, coffee, bananas, and other new crops were introduced, many native habitats like mangrove forests were destroyed.

When Europeans began to control Puerto Rico, there may have been 12,146 hectares (30,000 acres) of mangroves on the island (Miller and Lugo 2009). By 1975, half had been destroyed due to agriculture and human developments along the coast. By the mid-1980s, up to 75% had been destroyed or highly altered. Endemic mangroves are among the most important plant communities for water filtration and protection against flooding. Mangroves dampen the power of storm waves, as their roots stabilize land that would otherwise be eroded into the sea. They filter nutrients and sediments, along with treating waste that is harmful to human health, and providing aquatic and avian habitat. Altered hydrology, impervious surfaces that do not allow water to be absorbed into the soil, and aging infrastructure all intensify vulnerability to weather extremes in cities. As Puerto Rico's industrial and service-dominated economies have expanded in recent decades, urban areas have rapidly developed in low-lying zones. These zones are more vulnerable to sea-level rise, increasing the need for ecosystems such as mangrove forests to stabilize soil and filter water.

During Hurricane María, major flooding occurred in coastal urban areas like the capital of San Juan (FEMA 2018), which is covered in sealed soil with little capacity to absorb stormwater. Most of San Juan's critical infrastructure, electric power plants, businesses, and hotels are located in the coastal zone of the San Juan Metropolitan Area (PRCCC 2013). As San Juan's aging sanitary and stormwater infrastructure were clogged with fallen debris and inundated by heavy rainfall, residents of low-lying La Perla and other neighborhoods waded through contaminated water to remove debris and transport people with medical emergencies to hospitals (Mazzeio and Martinez 2017).

2.2.1 Urban Industrialization

Urban industrialization is a process through which greenhouse gas emissions changed the climate at a global scale, but which also created unique material and social vulnerabilities in the very sites of that production. The Industrial Revolution—which began in England between 1760 and 1840, and spread throughout Europe and then North American during the early nineteenth century—marked the shift to fossil fuel-based energy. Coal fueled expansion of the steam engine and transportation, as well as the dominance of the factory system of production over home labor and agriculture. This shift from home labor and agriculture to factory production centralized laborers in cities. Prior to the industrial revolution in England, a series of economic shocks and the practice of enclosure had pushed tenant farmers off their land and driven households into economic decline. Poor relief was provided to supplement wages, based on the price of bread and the size of the family. As the demand for urban labor grew, these relief programs became controversial, and the state was blamed for creating indigency by interfering in self-regulating labor markets (Block and Somers 2003). Thus, the Poor Law Amendments—which required the indigent to receive "assistance" in urban workhouses—were passed in 1834. Within 20 years, the population of cities had doubled and urban communities underwent major physical transformations as the need for housing, sanitary water and sewer, transportation, and other infrastructure became concentrated alongside these industrial activities.

2.2.2 Urban Climatology

Historical processes of urbanization—including urban density and expansion—have not only warmed the globe through greenhouse gas-emitting activities but have also warmed the local climate and changed local weather patterns through land use changes. Cities replace agricultural and rural landscapes with mineral-based materials that seal soil and warm the environment (Stone 2012). Sealed soil leads to more stormwater run-off and reduces the land's natural ability to cool itself. Engineered materials also alter the albedo or reflectivity of the land surface; cities absorb 80–85% of incoming solar radiation, making them hotter than non-urban locations (Taha 1997). In addition, industrial and transportation activities produce waste heat emissions. Where human-engineered materials predominate, heat waves, precipitation events, droughts, and wildfires are more intense since heat provides fuel for storms and dries the air mass. The increasing concentration of people in cities where these materials predominate makes cities particularly risky places to live in the context of climate change.

2.3 Breaking Political Will

Although extreme weather events (such as heavy precipitation or high temperature) are environmental phenomena, these phenomena and associated human vulnerabilities are altered and exacerbated by social processes that are political, economic, and exclusionary in nature. Industrial, transportation, and residential land use-driven greenhouse gas emissions—which make extreme events more likely and exacerbate their impacts—are the result of political action and inaction. Political agendas entail favoritism for certain types of economic activity and power over others, and use rhetoric ranging from economic liberty to stigmatization of the communities which threaten and are oppressed in order to maintain that power. Altering the climate is an economic and political endeavor, as is curbing climate change and adapting to climate-exacerbated extreme weather events.

At a global scale, inter-jurisdictional relationships of trade and governance play important roles in climate (in)action. Scientific, economic, and political narratives obscure the political economy drivers of climate change and its relation to class, race, and other means of oppression. Particularly for places like Puerto Rico, colonial histories and political status that convey rights to particular types of claims and not others play important roles in recovering from and building resilience to extreme weather events.

Prior to the U.S. invasion of Puerto Rico during the 1898 Spanish-American War, the island had just won autonomy from Spain, enabling it to elect voting delegates to the Spanish Parliament and its own legislature. The American military invaded with promises to help secure Puerto Rico's liberty, motivating many Puerto Ricans to fight against the Spanish during this invasion. That promise of liberty was not upheld. Once the Treaty of Paris was signed, the United States instituted a system of colonial rule over Puerto Rico, and made issues of citizenship ambiguous (Erman 2018). Under the Foraker Act of 1900, the governor of Puerto Rico and all other major offices were appointed by the U.S. President, and local laws were subject to veto by U.S. Congress (Miller and Lugo 2009). Puerto Rico was prohibited from negotiating trade treaties with foreign nations and the peso was retired at 60 cents to the dollar, causing a 40% rise in the cost of living and a fall in the price of land, enabling U.S. sugar corporations to purchase vast areas for sugar production.

Until President Woodrow Wilson signed the Jones-Shafroth Act in 1917 so that Puerto Ricans could enlist and be drafted to WWI, the people of Puerto Rico had limited legal rights. Jones-Shafroth granted Puerto Ricans Congressional citizenship but not Constitutional citizenship, though the Act was eventually amended in 1947 to allow governors to be popularly elected. Additionally, the island would become its own Commonwealth and develop its own Constitution in 1952. However, this Commonwealth status came with neither full independence nor full U.S. citizenship rights. Today, Puerto Ricans still do not elect voting representation to Congress or vote in presidential elections unless they reside in one of the 50 states.

Following Hurricane María, U.S. President Trump characterized Puerto Ricans as lazy and threatened to reduce recovery efforts, claiming on Twitter that "they want everything done for them." Stigmatization, misrecognition, and other status injuries form important means through which distributional and procedural inequities are institutionalized. Without representation in Congress or the ability to vote in general elections, Puerto Rican residents and their leaders have little political leverage to rapidly direct critical emergency resources. Indeed, recovery resources for Puerto Rico were insufficient and withheld for long periods. Here, we see a form of procedural inequity reinforced through status injury and translating into a material withholding, or distributional inequity. By comparison, recovery benefits to victims of Hurricane Harvey—which struck Houston, Texas only weeks earlier than Hurricane María struck Puerto Rico—were more generous and quickly deployed. A Politico Investigation outlined these discrepancies:

> Within six days of Hurricane Harvey, U.S. Northern Command had deployed 73 helicopters over Houston, which are critical for saving victims and delivering emergency supplies. It took at least three weeks after María before it had more than 70 helicopters flying above Puerto Rico.
>
> Nine days after the respective hurricanes, FEMA had approved $141.8 million in individual assistance to Harvey victims, versus just $6.2 million for María victims.

> During the first nine days after Harvey, FEMA provided 5.1 million meals, 4.5 million liters of water and over 20,000 tarps to Houston; but in the same period, it delivered just 1.6 million meals, 2.8 million liters of water and roughly 5,000 tarps to Puerto Rico.
>
> Nine days after Harvey, the federal government had 30,000 personnel in the Houston region, compared with 10,000 at the same point after María.
>
> It took just 10 days for FEMA to approve permanent disaster work for Texas, compared with 43 days for Puerto Rico.
>
> Seventy-eight days after each hurricane, FEMA had approved 39% of federal applications for relief from victims of Harvey, versus 28% for María (Vinik 2018).

Comparing measures of federal spending, federal resources distributed, and direct and indirect storm-mortality estimates, Willison et al. (2019) found that although Hurricane María caused more damage in Puerto Rico than did Hurricane Irma in Florida or Harvey in Texas, fewer resources were dedicated to the post-storm recovery in Puerto Rico.

2.3.1 *Liberal Trade Narrative*

During the 1980s, scientific consensus was beginning to form an understanding that climate change is anthropogenically-driven. Klein (2014) argues that the timing of this consensus-forming unfortunately coincided with the adoption of major trade deals that made climate action subordinate to the liberal trade agenda. In the early 1990s when global leaders were negotiating how to reduce greenhouse gas emissions through the United Nations Framework Convention on Climate Change (UNFCCC), the World Trade Organization (WTO) adopted a set of policies that undermined the UNFCCC. Free trade agreements governed by the WTO limit national economic sovereignty, and in some cases have stymied communities' abilities to locally invest in green energy technology. Rhetoric against these programs—which involve subsidies and local industry requirements—claim they distort the free market; yet, fossil fuel companies receive up to US $1 trillion in annual subsidies (Bast et al. 2012), limiting the extent to which green energy technology can freely compete. Political economy narratives such as those associated with liberal trade have been particularly effective at obstructing collective climate action that would curb carbon-producing activities.

2.3.2 *Rational Choice Narrative*

Economic narratives of climate change as a market failure or market externality are equally problematic, as they position climate change as a phenomenon that can be dealt with via market correction rather than as a result and necessary outcome of the growth-based global economy (Daly 1996). Rational choice economic theory assumes that people make decisions by choosing among possible alternatives an option that maximizes their utility based on preferences and budget constraints. Under this ideology, behavior and decisions are themselves evidence of people's preferred choice. Rather than placing responsibility on collective action of societies (e.g., on governments) and society, sustainability is cast as a matter of individual choice to be achieved via market-based approaches that incentivize pollution reduction and shift consumption patterns in favor of green products. This rational choice-based sustainability narrative obscures the far more impactful role of institutions and systems in producing climate change (and influencing individual choices (Simon 1986)). For instance, if the majority of transportation dollars are spent on optimizing the number of automobiles that can travel along our highway system, what actual role does personal choice afford in making more sustainable mobility decisions?

Similar to liberal trade narratives—which obscure the role of government in providing subsidies to support fossil fuel industries—the rational choice narrative has associated liberty and freedom of choice values with automobile-based transportation in order to obscure the role of government subsidy. Peter Norton (2015) argues that the automobile industry and U.S. government used a "Love Affair Thesis" to convince the Americans that they prefer automobile-based transportation in spite of its safety risks. In the United States, 43–74% of the cost of the highway system has been directly subsidized by government rather than supported through gas taxes (Dutzik et al. 2015). Yet, the personal automobile is cast as a "private" form of transportation in which individual-level rational choice plays the dominant role in behavioral decisions compared with mass, or "public" transit which the government coerces people to pay for and use. Thus, political economy narratives such as those associated with rational choice theory (like those associated with liberal trade) are effective at stymying collective action to change carbon-producing behaviors such as those in the transportation sector.

2.3.3 Global Climate Narrative

While scientific narratives may be designed to encourage particular types of climate action in a global arena, they effectively obscure more granular distributional dimensions of climate change and extreme event impacts in ways that reinforce power relations. In 2016, parties to the United Nations Framework Convention on Climate Change (UNFCCC) Paris Agreement agreed to limit global temperature rise to below 2 °C, or about 3.6°F. However, given that a global temperature rise of 2 °C will entail major rises in sea level, longer droughts and more intense heat waves among other effects (not to mention that communities across the globe have already experienced irreversible impacts from climate change), this threshold is more political than scientific. Shaw (2015) argues that "2-degrees" is a symbol which "validates stories that are designed to mask the conflicts between interests of the ruling class and those of society as a whole" (p. 10). Global averages obscure the uneven geographies of responsibility—who is benefiting and who is bearing the burden of climate change. Although scientific narratives focus on estimates of average historic and projected warming, the globe is not warming evenly and people (at the national, sub-national, or even neighborhood level) do not have equal capacities to cope with warming experienced in their communities. Over the past half century, it is estimated that climate change has increased between-country inequality by 25%, associated with economic output declines in hotter poorer countries, and economic growth in cooler, wealthier countries (Diffenbaugh and Burke 2019). Places like Puerto Rico are much more vulnerable to climate change, not only due to their coastal location but also because they lack political leverage and political autonomy to harness resources that would enable climate change-related mitigation, adaptation, and response. Although greenhouse gases accumulate in the atmosphere globally, not all parts of the globe experience this accumulation in the same way.

2.4 Urban Climate Extremes Exacerbate Existing Inequities

Social separateness embedded in many institutional forms is exacerbated by climate change. In the case of Puerto Rico's sovereign political status, we saw that distributional inequities are reinforced through procedural inequities and misrecognition. But distributional inequities also exist between different communities within territorial boundaries and cities. These are codified in exclusionary practices of misrecognition and unfair participation that determine, for instance, who has access to protective housing and financial resources, or even whose fatalities are counted among the climate-induced.

Calle Norzagaray in Old San Juan overlooks the Atlantic Ocean. An elevated route used by tourists to visit the historic Castillo de San Cristóbal also forms a boundary with the low-lying historic shantytown neighborhood of La Perla. Along this panorama of economic and climate injustice is a mural which refutes the U.S. government's official Hurricane María death count (Fig. 2.2).

There are multiple ways to calculate death rates due to extreme weather. Conventionally, the practice is to use death certificate reporting. Known as *directly related disaster death*, this accounts for people who die during the peak of the storm from physical impacts like structural collapse. According to the Centers for Disease Control and Prevention, *indirectly related disaster death* should also be reported. This type of fatality could be due to unsafe or unhealthy conditions during pre-event preparations or post-event cleanup (US DHHS 2017). However, even this approach does not account for indirect deaths due to longer term loss of water, emergency services, and delays or interruptions in health care due to power outages. Kishore et al. (2018) conducted a stratified survey-based study of 3,299 households from January 17 to February 24, 2018 to compare death rates during that period to the same period in the prior year. Using a measure of "excess mortality," the researchers estimated 4,645 deaths likely due to Hurricane María, far higher than the official U.S. government count of 64. The mural on Calle Norzagaray draws attention to this discrepancy and the broader misrecognition of the very nature of hurricane-related

Fig. 2.2 "Recordemos nuestros Muertos (Remember our dead) 4,645" mural overlooking La Perla, Old San Juan, Puerto Rico. Image credit: the author

vulnerability. Extreme weather-related deaths are not mere acts of nature. Recognizing that lack of critical infrastructure and services are to blame for thousands of deaths would implicate people and institutions, especially those with power.

Whereas conventions used for causal attribution only allow classification of deaths attributable to immediate physical impact, excess mortality captures the fragility of our infrastructure and service provisioning in the face of extreme events, particularly in ways that impact low-income communities of color who may not be able to afford energy generators or flights to safe communities (e.g., the U.S. mainland). Moreover, spatial segregation along racial and class lines often places low-income people of color in communities most exposed to environmental burdens, including extreme weather events.

Inequity is produced at the intra-urban scale through institutions that govern access to jobs, housing, transportation, and land use. Early industrialization formed cities, but those cities were reconfigured during industrial relocation processes that occurred during the middle of the last century, along with suburbanization processes which segregated poor communities of color in inner cities that lacked job opportunities. These reconfigurations reshaped both the biophysical and social contexts of cities in ways that would ultimately—due to a coupling of land use and social segregation—concentrate poor communities of color in hot microclimates and locations that are disproportionately exposed to environmental burdens such as air pollution, toxic waste, and flooding (Bullard 2000).

Throughout the twentieth century in the United States, an enormous wealth gap grew between black and white households. In 1935, the Home Owners Loan Corporation (HOLC) was established to insure home mortgages following the banking crisis of the 1930s that had resulted in many foreclosures. Favoring suburban development for white households, this insurance was unavailable to racial and ethnic minorities, immigrants and for homes situated in urban communities. It was the HOLC's policy to ensure that "incompatible racial groups should not be permitted to live in the same communities" (Federal Housing Administration 1936). This policy was codified through residential security "redlining" maps, which delineated boundaries around communities in which it was considered "safe" or "unsafe" to offer insurance. As cartographic guide to a host of local land use tactics, residential security maps created a spatial logic by which officials, developers, and real estate agents would: (a) create suburban all-white neighborhoods, (b) rigidize existing segregation patterns, and (c) segregate formerly integrated neighborhoods. As white neighborhoods were rezoned as single-family only, black neighborhoods were rezoned commercial or industrial (or granted variances for such uses) (Rothstein 2017). Under the auspices of economic protection, uses like polluting industries could only exist in black neighborhoods. Recent research finds that surface temperatures in redlined communities are 2.6 °C (4.7°F) hotter relative to non-redlined neighboring communities (Hoffman et al. 2020). In addition to inscribing environmental burden like heat in the built environment, redlining also created unequal access to wealth. Since home ownership became affordable for white families, it formed a primary means through which middle class families generate wealth and pass on wealth to their children in the form of higher education and other investments. In addition, residential property tax forms an important means through which communities collectively invest in local infrastructure. Household and neighborhood-level resources can be crucial for resilience to heat or flooding, as access to quality weatherized housing, energy for air conditioning, and flood insurance is largely a matter of income and wealth. Communities already facing high food, energy, and housing burdens have greater sensitivities to extreme weather events.

Scholars of racial capitalism have argued that racism and capitalism are intrinsically linked, since capitalist wealth production is not possible without the production of social separateness that can be used as a tool to exploit labor (Melamed 2015; Pulido 2017). Social separateness is embedded in processes of wealth accumulation and economic development that determine land use patterns, and it produces residential locations, the hazards to which people living in those residential locations are exposed, and spaces that enhance or minimize collective life. Moreover, extreme weather exacerbates forms of separateness which are ingrained in unfair institutions of housing, transportation systems, and economy that often predate an extreme event. In Gulf region states of the United States affected by Hurricane Katrina, women and women of color were most likely to lack access to health insurance, experience disproportionately high rates of poverty, and engage in low-wage work despite high work participation rates (Jones-DeWeever and Hartman 2006). These forms of exclusion that were in place prior to the storm made relocation and recovery exceptionally arduous after the storm. Understanding how historical processes of industrial extraction, as well as housing, transportation, and other land use policies and practices have formed spatial configurations across cities is crucial for recognizing and mitigating contemporary forms of exclusion and the production of vulnerability to weather extremes.

2.5 Conclusion

Our climate and humanitarian crisis has been produced by an economic system that devalues labor and natural resources, and which is supported by housing, transportation, and land use institutions that reinforce vulnerability of oppressed groups through exclusionary practices. At the same time, barriers to addressing these crises are cast as a problem of individual-level choice that could be rectified by market-based approaches. The Human Rights Council of the United Nations has used the term "climate apartheid" to describe the impacts of an overreliance on the private sector for climate disaster protection (United Nations Human Rights Council 2019). Addressing these crises must involve transparency about who is benefiting, who is burdened, and how we institutionally produce this inequality. Scientific narratives that generalize climate change signals and impacts based on averages without also describing distributional exposures and disproportionate impacts exacerbate this inequality. Examining how social separateness that forms unequal climate burden is embedded in planning, policy, economic, and scientific practices is crucial for transforming those institutions to design adaptation strategies that protect vulnerable communities from extreme weather.

Creating resilient urban futures will involve forecasting, projecting, and visioning, using all the technical and creative tools at our disposal. However, we cannot truly understand the nature of the problems we are trying to solve for the sake of future generations unless we understand how problems were produced in the past. This is because climate change and vulnerabilities to weather extremes have been institutionalized in structures that are designed to replicate themselves. Resilient futures practice is a multi-sectoral activity, as we see that vulnerability is embedded not only in systems of ecology and health, but also economy and housing. Each of these and many more sectors has played a role in producing oppression and each must reckon with that history in developing strategies for transforming social separateness into social cohesion. Environmental justice theory offers conceptual tools for unpacking ways in which misrecognition, distributional injustice, procedural injustice, and other forms of social separateness exist in such institutions. In order to create positive and transformative future narratives that will show us the way forward, we must first construct narratives that help us understand where we are and how we got here. Thus, urban futures work situated in history and informed by environmental justice theory is an essential first step toward climate equity and resilience.

Acknowledgements This work was supported by the Urban Resilience to Extreme Weather-Related Events Sustainability Research Network (UREx SRN; NSF grant no. SES 1444755). The author is grateful to Joyce Coffee, Philip Gilbertson, Elizabeth Cook, and to Tico Taussig for their thoughtful comments.

References

Bast E, Kretzmann S, Krishnaswamy S et al (2012) Low hanging fruit: Fossil fuel subsidies, climate finance, and sustainable development. Heinrich-Böll-Stiftung, Washington, D.C. https://priceofoil.org/content/uploads/2012/06/LowHangingfruit.pdf. Accessed 15 Jun 2020

Block F, Somers M (2003) In the shadow of speenhamland: social policy and the old poor law. Polit Soc 31:283–323. https://doi.org/10.1177/0032329203252272

Bullard RD (2000) Dumping in Dixie: race, class and environmental quality, 3rd edn. Westview Press, Boulder CO

Daly H (1996) Beyond growth: the economics of sustainable development. Beacon Press, Boston, MA

Diffenbaugh NS, Burke M (2019) Global warming has increased global economic inequality. Proc Natl Acad Sci 116(20):201816020. https://doi.org/10.1073/pnas.1816020116

Dornbusch R (1985) Purchasing power parity. Nat Bur Econ Res (1591). https://doi.org/10.1007/3-540-68921-4_10

Dutzik T, Weissman G, Baxandall P (2015) Who pays for roads? How the "users pay" myth gets in the way of solving America's transportation problems. United States Public Interest Research Group (US PIRG) Education Fund, Frontier Group, p 1–40.

Eckstein D, Hutfils ML, Winges M (2019) Global climate risk index 2019: who suffers most from extreme weather events? Weather-related loss events in 2017 and 1998 to 2017. Germanwatch, Bonn, Germany

Erman S (2018) Almost citizens: Puerto Rico, the U.S. Constitution, and empire. Cambridge University Press, Cambridge, UK

Federal Emergency Management Commission (FEMA) (2018) Hurricanes Irma and María in Puerto Rico: Building performance, observations, recommendations and technical guidance. Available via Homeland Security Digital Library. https://www.hsdl.org/?view&did=818977. Accessed 15 Jun 2020

Federal Housing Administration (1936) Underwriting manual: underwriting and valuation procedure under title II of the national housing act. Available via Office of Policy Development and Research. https://www.huduser.gov/portal/sites/default/files/pdf/Federal-Housing-Administration-Underwriting-Manual.pdf. Accessed 16 Jun 2020

Ferré-Sadurní L (2017) In a Puerto Rican Town, "Water Came Out of Nowhere." New York times. https://www.nytimes.com/2017/09/22/us/puerto-rico-toa-baja-hurricane-.html. Accessed 15 Jun 2020

Funes Y (2017) Puerto Rico had towering landfills and coal ash pollution. Then, María Hit. Grist. https://grist.org/article/puerto-rico-had-towering-landfills-and-coal-ash-pollution-then-Maria-hit/. Accessed 16 Jun 2020

Gould WA, Díaz NL, Álvarez-Berríos F et al (2018) U.S. Caribbean. In: Reidmiller DR, Avery CW, Easterling DR et al (eds) Impacts, risks, and adaptation in the United States: fourth national climate assessment, vol. II. U.S. Global Change Research Program, Washington, D.C. https://doi.org/10.7930/NCA4.2018.CH20

Hamstead Z, Coseo P (2020) Transforming global health. In: Smith KH, Ram PK (eds) Transforming global health. Springer Nature Switzerland, Cham, Switzerland, pp 261–283. https://doi.org/10.1007/978-3-030-32112-3

Hinojosa J, Meléndez E (2018) Puerto Rican exodus: One year since Hurricane María. Center for Puerto Rican Studies, Hunter College, City University of New York. https://centropr.hunter.cuny.edu/research/data-center/research-briefs/puerto-rican-exodus-one-year-hurricane-Maria. Accessed 16 Jun 2020
Hoffman JS, Shandas V, Pendleton N (2020) The effects of historical housing policies on resident urban areas. Climate 8(1):1–15. https://doi.org/10.3390/cli8010012
Intergovernmental Panel on Climate Change (IPCC) (2007) Food, fibre and forest products. In: Parry M, Canziani O, Palutikof J et al (eds) Climate change 2007: impacts, adaptation and vulnerability. Contribution of Working Group II to the Fourth Assessment Report of the Intergovernmental Panel on Climate Change. Cambridge University Press, Cambridge, UK
Intergovernmental Panel on Climate Change (IPCC) (2012) Glossary of terms. In: Field CB, Barros V, Stocker TF et al (eds) Managing the risks of extreme events and disasters to advance climate change adaptation. Cambridge University Press, Cambridge University Press, Cambridge, UK, and New York, NY, pp 555–564
Intergovernmental Panel on Climate Change (IPCC) (2014) Climate Change 2014: Synthesis Report. Contribution of Working Groups I, II and III to the Fifth Assessment Report of the Intergovernmental Panel on Climate Change. IPCC, Geneva, Switzerland, p 151
Jones-DeWeever AA, Hartman H (2006) Abandoned before the storms: the glaring disaster of gender, race and class disparities in the Gulf. In: Squires G, Hartman C (eds) There is no such thing as a natural disaster: race, class and Hurricane Katrina. Routledge, New York, NY, pp 85–102
Kishore N, Marqués D, Mahmud A et al (2018) Mortality in Puerto Rico after Hurricane María. N Engl J Med 379(2):162–170. https://doi.org/10.1056/NEJMsa1803972
Klein N (2014) Hot money: how free market fundamentalism helped overheat the planet. This changes everything: capitalism vs the climate. Simon & Schuster, New York, NY, pp 64–95
Mazzeio P, Martinez E (2017) As flooding continues across Puerto Rico, San Juan begins to pick up after Hurricane María. Miami Herald. https://www.miamiherald.com/news/weather/hurricane/article174640821.html. Accessed 21 Jun 2020
Melamed J (2015) Racial Capitalism. Crit Ethn Stud 1(1):76–85. https://doi.org/10.5749/jcritethnstud.1.1.0076
Miller GL, Lugo AE (2009) Guide to the ecological systems of Puerto Rico. USDA Gen Tech Rep 35:1–436. https://doi.org/10.2737/IITF-GTR-35
Norton P (2015) Of love affairs and other stories. Incomplete streets: processes, practices, and possibilities. Routledge, London, pp 17–35
Puerto Rico Climate Change Council (PRCCC) (2013) Working group 3 report: Climate change and Puerto Rico's society and economy. In: Jacobs KR, Terando A, Diaz E (eds) Puerto Rico's State of the Climate 2010–2013. Puerto Rico Coastal Zone Management Program, Department of Natural and Environmental Resources, NOAA Office of Ocean and Coastal Resource Management, San Juan, Puerto Rico
Pulido L (2017) Geographies of race and ethnicity II: environmental racism, racial capitalism and state-sanctioned violence. Prog Hum Geogr 41(4):524–533. https://doi.org/10.1177/0309132516646495
Rothstein R (2017) The color of law: a forgotten history of how our government segregated America. W.W. Norton & Company Inc., New York, NY
Schlosberg D (2007) Defining environmental justice: theories, movements, and nature. Oxford University Press, Oxford, UK
Seneviratne S, Nicholls N, Easterling D et al (2012) Changes in climate extremes and their impacts on the natural physical environment. In: Field CB, Barros V, Stocker TF et al (eds) Managing the risk of extreme events and disasters to advance climate change adaptation. A Special Report of Working Groups I and II of the Intergovernmental Panel on Climate Change (IPCC). Cambridge University Press, Cambridge, UK and New York, NY, pp 109–230. https://doi.org/10.2134/jeq2008.0015br

Shaw C (2015) The two degrees dangerous limit for climate change: Public understanding and decision making. Routledge, Abingdon, UK
Simon HA (1986) Rationality in psychology and economics. J Bus 59(4):S209–S224. https://doi.org/10.1086/296363
Stone B Jr (2012) The city and the coming climate: climate change in the places we live. Cambridge University Press, New York, NY
Taha H (1997) Modeling the impacts of large-scale albedo changes on ozone air quality in the South Coast Air Basin. Atmos Environ 31(11):1667–1676. https://doi.org/10.1016/S1352-2310(96)00336-6
United States Department of Health and Human Services (US DHHS) Centers for Disease Control and Prevention, National Center for Health Statistics (2017) A reference guide for certification of deaths in the event of a natural, human-induced, or chemical/radiological disaster. U.S Department of Health and Human Services, Washington, DC
United Nations Human Rights Council (2019) Climate change and poverty: Report of the Special Rapporteur on extreme poverty and human rights. United Nations Digital Library. https://www.file:///C:/Users/galac/AppData/Local/Packages/Microsoft.MicrosoftEdge_8wekyb3d8bbwe/TempState/Downloads/A_HRC_41_39-EN.pdf. Accessed 21 Jun 2020
USGCRP (2016) Disruption of essential infrastructure. In: Crimmins A, Balbus J, Gamble JL et al (eds) The impacts of climate change on human health in the United States: A scientific assessment. U.S. Global Change Research Program, Washington, D.C. https://doi.org/10.7930/J0R49NQX
Vinik D (2018) How Trump favored Texas over Puerto Rico. Politico. https://www.politico.com/story/2018/03/27/donald-trump-fema-hurricane-Maria-response-480557. Accessed 21 Jun 2020
Willison CE, Singer PM, Creary MS et al (2019) Quantifying inequities in US federal response to hurricane disaster in Texas and Florida compared with Puerto Rico. BMJ Glob Heal 4(1):1–6. https://doi.org/10.1136/bmjgh-2018-001191
World Health Organization (2018) Climate change and health. World Health Organization. https://www.who.int/news-room/fact-sheets/detail/climate-change-and-health. Accessed 18 July 2018
Yonetani M, Albuja S, Bilak A et al (2015) Global estimates 2015: people displaced by disasters. International Displacement Monitoring Centre, Geneva, Switzerland

Zoé A. Hamstead is an Assistant Professor of Environmental Planning at the University at Buffalo and Director of the Community Resilience Lab. Her research builds on interdisciplinary approaches in urban planning, geography, and urban ecology to support climate equity and adaptive climate planning approaches. Dr. Hamstead teaches courses in environmental planning, environmental justice, economic concepts, geographic analysis of environmental problems, and engaged community energy studios. She holds a Ph.D. in Urban & Public Policy from The New School, a Master's in City & Regional Planning from the University of North Carolina at Chapel Hill, and a Bachelor of Liberal Arts from St. John's College.

Chapter 3
Social, Ecological, and Technological Strategies for Climate Adaptation

Yeowon Kim, Lelani M. Mannetti, David M. Iwaniec, Nancy B. Grimm, Marta Berbés-Blázquez, and Samuel Markolf

Abstract Resilient cities are able to persist, grow, and even transform while keeping their essential identities in the face of external forces like climate change, which threatens lives, livelihoods, and the structures and processes of the urban environment (United Nations Office for Disaster Risk Reduction, How to make cities more resilient: a handbook for local government leaders. Switzerland, Geneva, 2017). Scenario development is a novel approach to visioning resilient futures for cities. As an instrument for synthesizing data and envisioning urban futures, scenarios combine diverse datasets such as biophysical models, stakeholder perspectives, and demographic information (Carpenter et al. Ecol Soc 20:10, 2015). As a tool to envision alternative futures, participatory scenario development explores, identifies, and evaluates potential outcomes and tradeoffs associated with the management of social–ecological change, incorporating multiple stakeholder's collaborative subjectivity (Galafassi et al. Ecol Soc 22:2, 2017). Understanding the current landscape of city planning and governance approaches is important in developing city-specific scenarios. In particular, assessing municipal planning strategies through the lens of interactive social–ecological–technological systems (SETS) provides useful insight into the dynamics and interrelationships of these coupled systems (da Silva et al. Sustain Dev 4(2):125–145, 2012). An assessment of existing municipal strategies can also be used to inform future adaptation scenarios and strategic plans addressing extreme weather events. With the scenario development process guiding stakeholders in generating goals and visions through participatory workshops, the content analysis

Y. Kim (✉) · S. Markolf
Julia Ann Wrigley Global Institute of Sustainability, Arizona State University, Tempe, AZ, USA
e-mail: yeowon.kim@asu.edu

L. M. Mannetti · D. M. Iwaniec
Andrew Young School of Policy Studies, Urban Studies Institute, Georgia State University, Atlanta, GA, USA

N. B. Grimm
School of Life Sciences, Arizona State University, Tempe, AZ, USA

M. Berbés-Blázquez
School for the Future of Innovation in Society, Arizona State University, Tempe, AZ, USA

Z. A. Hamstead et al. (eds.), *Resilient Urban Futures*, The Urban Book Series,
https://doi.org/10.1007/978-3-030-63131-4_3

of governance planning documents from the SETS perspective provides key insight on specific strategies that have been considered (or overlooked) in cities. In this chapter, we (a) demonstrate an approach to examine how cities define and prioritize climate adaptation strategies in their governance planning documents, (b) examine how governance strategies address current and future climate vulnerabilities as exemplified by nine cities in North and Latin America where we conducted a content analysis of municipal planning documents, and (c) suggest a codebook to explore the diverse SETS strategies proposed to address climate challenges—specifically related to extreme weather events such as heat, drought, and flooding.

Keywords Climate adaptation strategies · Governance · Resilience planning · Social-ecological-technological system (SETS)

3.1 Social–Ecological–Technological Systems (SETS) Framework

Envisioning how we transform our cities into places and communities that are resilient is an emerging urban challenge that requires an approach integrating diverse knowledge, experience, and perspectives (Muñoz-Erickson et al. 2017). Cities are SETS, and so are parts of cities like neighborhoods, parks, and various types of infrastructure. The SETS perspective is an important aspect of envisioning urban futures because cities are considered as systems, meaning we cannot consider parts of cities—institutions, ecosystems, built environment, and communities—in isolation since they interact to form the whole.

In SETS, social dimensions include social–political–cultural–economic dynamics of a city, including both the decision-making actors and their actions. Ecological dimensions include the biophysical elements of non-human nature, with their associated processes, that are part of the fabric of cities—for example, tree growth or soil formation. Technological dimensions include the built components and associated processes of urban systems, for example, roads or public transportation networks, buildings, and the knowledge embodied in technologies (Markolf et al. 2018). Envisioning cities from a SETS perspective raises valuable governance questions, such as the type of institutions and knowledge needed, as well as which people are affected by infrastructure changes (Kim et al. 2019). How can services provided by natural ecosystems be integrated into the built environment? How can technological advances be used to impart flexibility or redundancy to infrastructure? The SETS approach demands that such questions—reflecting the three SETS dimensions—be answered to build resilience and support sustainable pathways.

The SETS framework for climate adaptation is a pragmatic approach that reflects an increasing recognition of the role that built and technological infrastructure play in mediating the relationships among human activities and ecosystem processes (Grimm et al. 2015; McPhearson et al. 2016). The SETS framework is fundamental to climate adaptation plans because it helps to clarify how interactions among the social–political–cultural–economic (S) and the biophysical (E) domains are mediated through infrastructure (T). Key SETS components to consider encompass diverse social, ecological, and technological features, as well as where these intersect, since these three dimensions interact with each other in supporting urban pathways to resilient futures. Examples include social–ecological considerations in land use changes, ecological effects of biophilia, or the need for more green spaces on society, and technological–social innovation for mobility or communication (Table 3.1).

3.2 Content Analysis of Municipal Planning Documents and Governance Strategies in SETS

In the face of the growing occurrence of weather extremes, climate adaptation plans are essential governance tools at regional, city, and local levels. Though such plans have been extensively developed at national and international levels, local governments have a vital role in implementing municipal-level climate adaptation strategies that are retrofit to various governance scales, regional climatic characteristics, and urban SETS. In the last two decades, city governments have been developing planning documents such as comprehensive municipal plans, disaster preparedness plans, climate action plans, and sustainability plans meant to advance urban resilience by implementing climate adaptation strategies at local levels (Reckien et al. 2018). City plans and city planning processes embody the goals and actions that cities seek to advance for urban resilience (Bulkeley 2010). Municipal governance is often shaped by various forms of interacting institutions, including governing agencies, policies, formal and informal codes, local knowledge systems, practitioners, public officials, and communities (Folke et al. 2005; Araos et al. 2016; Feagan et al. 2019). City plans express goals that are shaped by the various institutions, as well as guide interactions among institutions to achieve goals, demonstrate suitable governance strategies, and envision achievable expectations and outcomes of these strategies (Carmin et al. 2012). As cities continue to lead urban resilience planning, we analyze municipal planning documents to examine how urban governance structures (with diverse socio–political–cultural and biophysical contexts) plan for climate change. Analyzing plans help us understand what strategies are effective and practical, and how well adaptation strategies are integrated in local governance. As such, governance planning documents provide insight into how cities are framing urban resilience, yet there are few mechanisms to effectively and efficiently highlight the suite of SETS climate adaptation strategies that cities are considering. In the following sections, we provide four essential steps for analyzing governance strategies from municipal planning documents by using the SETS framework in order to support an effective scenario-development process for visioning resilient urban futures.

Table 3.1 Matrix exemplifying SETS features in cities. Here the social, ecological, and technological (S-E-T) domain characteristics (vertical header column) impact/influence the social, ecological, and technological components (horizontal header row) of a city. For example, the bottom-leftmost box indicates the ways in which technology influences society

SETS Domain	Social component	Ecological component	Technological component
Social (S)	Equity; Funding; Public education/engagement; Policy; Culture; Legislation; Public perception; Motivation	Land use (e.g., developed space vs. green space); Land conservation; Environmental advocacy groups; Community environmental action; Environmental regulations	Design standards/codes/plans; Institutional interactions; Standardization; Investment; Economic/social pressures
Ecological (E)	Biophilia; Public parks and preserves; Ecosystem services (e.g., clean air, clean water, etc.)	Habitat; Ecosystem; Water bodies; River systems; Forests; Natural resources	Environmental facilities; Environmental management technologies (e.g., stormwater and sewer systems); Fuel system inputs
Technological (T)	Mobility; Economic activity/opportunity; Communication; Comfort/protection from climate	Pollution (air, water, soil); Resource consumption; Monitoring of ecosystem health; Restoration of degraded ecosystems	Infrastructure; Maintenance/management/upgrades/replacement; Data availability and quality; Modelling

3.2.1 Selecting Municipal Planning Documents

The first step is to choose appropriate documents for analysis. Since our focus is municipal governance strategies for climate change adaptation, the pool of potential documents for analysis is limited to plans that are drafted and published by the city, local, and regional governments, and by local non-governmental organizations. Once the potential documents are identified in a city, three to five dominant governance documents are selected for analysis based on the following criteria.

- Must be an overarching planning document (e.g., General Plans, Comprehensive Plans, Sustainability/Resilience Plans, Climate Action Plans, Common Plans)
- Must be less than five years old, with exceptions if the total number of available documents for analysis in a city is less than three
- Must be relevant to climate change, flooding/heat/drought adaptation, resilience, or sustainability
- If more than five documents are available that fit the above criteria, only those salient to climate change adaptation, sustainability, or resilience are selected. If the document is titled with climate action, sustainability, or resilience, it may be prioritized, otherwise the relevance may be determined by how comprehensively the document focuses on strategic planning for mitigation of climatic risks or adaptation to environmental changes (e.g., comprehensive municipal plans, hazard mitigation plans, disaster preparedness management plans, stormwater plans)
- Match the plans to the spatial scale under consideration (e.g., neighborhood, city-wide, regional, national).

We recommend consultation and validation with city practitioners regarding the priority and relevance of documents to finalize the selection. Using the above section criteria, 30 planning documents from across the UREx SRN cities were selected for analysis. These include a diversity of document types relevant to climate adaptation, resilience, and sustainability. The selected documents were published between 2010 and 2015 at the municipal, regional, and state levels (Table 3.2).

3.2.2 Extracting Governance Strategies

From the selection of municipal plans in each city, governance strategies are extracted by capturing exact quotes from documents. The extraction should focus primarily on quotes that describe implementation strategies relating to extreme weather events (namely flooding, extreme heat, and drought), actions, or approaches to adapt to climate change or extreme events in general, and governance mechanisms to mitigate, adapt, or respond to events related to climate change. Examples of strategies extracted from across the UREx SRN cities are presented in Table 3.3.

Table 3.2 List of municipal planning documents selected for content analysis of governance strategies among the nine UREx network cities. Each document reflects climate adaptation, sustainability, or resilience

City	Governance level	Date published	Document name
Baltimore	Municipal	2006 (updated 2009)	Comprehensive Municipal Plan
Baltimore	Municipal	2013	Disaster Preparedness and Planning Project
Baltimore	Municipal	2013	Baltimore Climate Action Plan
Baltimore	Municipal	2009	The Baltimore Sustainability Plan
Hermosillo	Municipal	2013–2014	Municipal Development Plan
Hermosillo	Municipal	2015	Plan de Acción Climática Municipal Hermosillo (2015) (PACMUN)/Municipal Climate Action Plan for Hermosillo
Hermosillo	Municipal	2010	Strategic Plan for Storm Sewers
Miami	Municipal	2013	Miami Comprehensive Neighborhood Plan
Miami	Regional	2012	Southeast Florida Regional Climate Action Plan
Miami	Regional	2010	Miami-Dade Green Print: Our Design for a Sustainable Future
New York City	Municipal	2015	One New York: The Plan for a Strong and Just City
New York City	Municipal	2013	PlaNYC: A Stronger, More Resilient New York
New York City	Municipal	2014	New York City Hazard Mitigation Plan
New York City	Municipal	2011 (updated 2014)	PlaNYC: A Greener, Greater New York
Phoenix	Municipal	2015	PlanPHX General Plan
Phoenix	Regional	2015	Multi-Jurisdictional Hazard Mitigation Plan
Phoenix	Municipal	2011	Water Resources Plan
Portland	Municipal	2012	The Portland Plan
Portland	Municipal	2015	Portland's Recommended Comprehensive Plan
Portland	Municipal	2014	Climate Change Preparation Strategy/Risk and Vulnerability Assessment

(continued)

Table 3.2 (continued)

City	Governance level	Date published	Document name
San Juan	Municipal	2003 (updated in 2012)	Territorial Ordinance/Municipal Land Use Plan (I and II)
San Juan	Municipal	2015	Comprehensive Mitigation Plan Update
San Juan	Regional	2015	San Juan Bay Estuary Plan
San Juan	State	2015	Puerto Rico Climate Change Council's Ruta Hacia La Resiliencia
Syracuse	Municipal	2012	City of Syracuse Comprehensive Plan 2040 (including Land Use and Development and Sustainability Chapters)
Syracuse	Regional	2012	Onondaga County Climate Action Plan
Syracuse	Regional	2010	Onondaga County Multi-Jurisdictional Hazard Mitigation Plan
Valdivia	Municipal	2010	Plan Regulador Comunal de Valdivia
Valdivia	Municipal	2015	Sustainable Valdivia: Plan of Action
Valdivia	Municipal	2012	Stormwater Master Plan

3.2.3 Labeling Strategies with Levers and Exogenous Drivers

After strategies are extracted, the individual strategies are first qualitatively coded for the type of climatic drivers being addressed (i.e., exogenous drivers) and the type of policy instruments being implemented (i.e., levers) (Lempert et al. 2003; Wiek and Iwaniec 2014; Iwaniec et al. 2020). In our case, climatic drivers refer to extreme weather events that impact cities, such as floods (urban, coastal, riverine, or non-specific), extreme heat, drought, and non-specific hazards. Policy instruments are governance mechanisms that may be manipulated to mitigate or respond to the impact of these drivers. Examples include research and plan development, intergovernmental coordination, maintenance of built infrastructure, economic incentives, and education and outreach.

Table 3.3 Example of extracted strategies found within the planning documents outlined in Table 3.2, demonstrating how governance strategies differ by document type and by city

City	Document name	Extracted strategy
Baltimore	Comprehensive Master Plan	"Restore and protect at least one mile per year of streams and river basins in floodplains and stream valley" (City of Baltimore 2009, p 139)
Hermosillo	Plan de Acción Climática Municipal Hermosillo 2015 (PACMUN)/Municipal Climate Action Plan for Hermosillo	Encourage planting of trees and expanding local flora (green areas): Implementation of native species when planting new trees and reducing the felling of trees on public roads [Translated from an original quote in Spanish] (p 86)
Miami	Southeast Florida Regional Climate Action Plan	"Review and assess current agricultural best management practices for the state of Florida for its management of projected climate impacts" (Southeast Florida Regional Climate Change Compact 2015, p AG-6)
New York City	One New York: The Plan for a Strong and Just City	"Expand public education efforts so that all New Yorkers know the risks they face during extreme weather events and other disasters" (City of New York 2015, p 225)
Phoenix	Multi-Jurisdictional Hazard Mitigation Plan	"Review existing general plan and zoning ordinance to determine how these documents help limit development in hazard areas" (Maricopa County 2015 p 367)
Portland	Portland's Recommended Comprehensive Plan	"Create a network of distinctive and attractive City Greenways that link centers, parks, schools, rivers, natural areas, and other key community destinations" (City of Portland 2020, p GP3-19)
San Juan	PRCCC's Ruta Hacia La Resiliencia	Develop green infrastructure plans that improve engineered coastal barriers [Translated from an original quote in Spanish, PRCCC 2015 p 79]

(continued)

Table 3.3 (continued)

City	Document name	Extracted strategy
Valdivia	Plan Regulador Comunal de Valdivia	Maintain or increase urban vegetation [Translated from an original quote in Spanish, Valdivia 2010 p 30]

3.2.4 The SETS Codebook

We developed the SETS codebook that helps us identify SETS components of governance strategies based on Denton et al. (2014), Berbés-Blázquez et al. (2017), Burch et al. (2017), and Iwaniec et al. (2020). The SETS codebook (Table 3.4) is developed in an inductive process by encompassing a pool of sample strategies and incorporating previous studies on systems governance analysis. We propose this codebook for analyzing governance strategies to be qualitatively coded by their contents and evaluated by the interaction of social, ecological, and technological domains. As a non-scale, system-level, bridging framework, this coding scheme allows cities and their stakeholders to explore SETS interaction and adaptation strategies associated with them in city to regional-level governance data. In Table 3.5, we include selected examples of governance strategies that are analyzed by the proposed SETS codebook. The outcome of the analysis creates a comprehensive framework to assess climate change adaptation strategies based on their synergies, conflicts, and tradeoffs across SETS domains.

3.3 Conclusion

In this chapter, we present an approach to identify and analyze municipal governance strategies using a SETS framework for urban resilience framework. Assessing governance strategies using a SETS framework is particularly valuable in the scenario-based visioning process. SETS governance strategies help stakeholders understand current dynamics of urban systems and explore adaptation options prioritized at various governance scales, and are thus useful for visioning futures when provided to diverse stakeholders in the process of developing participatory scenarios. Analysis of governance strategies using a SETS framework can explain how cities currently address climate risks and existing system vulnerabilities through governance adaptation mechanisms. We are particularly interested in determining whether planning documents tend to prioritize a particular SETS domain over others (e.g., predominance of technological solutions), and if they adequately consider system relationships. Identifying SETS interactions in proposed and implemented municipal governance plans is an important step in bridging the gap between aspirations and

Table 3.4 The SETS codebook developed to capture SETS components of governance mechanisms in strategies

SETS domain	SETS code	SETS component	Strategies exemplifying component
Social	S1	Social safety nets	Social safety nets and social protection, food banks and distribution of food surplus, municipal services (including water and sanitation), vaccination programs, essential public health services (including reproductive health services), enhanced emergency medical services
	S2	Educational	Awareness raising and integrating into education, gender equity in education, extension services, sharing local and traditional knowledge, integration of local and traditional knowledge into adaptation planning, participatory action research and social learning, community surveys, knowledge-sharing and learning platforms, international conferences and research networks, communication through media, operations training. *S2 includes any type of knowledge transfer to stakeholders delineated within a strategy
	S3	Informational	Hazard and vulnerability mapping, early warning and response systems, systematic monitoring and remote sensing, climate forecast services, downscaling climate scenarios, longitudinal datasets, integrating indigenous climate observations, community-based adaptation plans (including community-driven slum upgrading and participatory scenario development). *S3 involves with data and information development
	S4	Behavioral	Household preparation and evacuation planning, retreat and migration, soil and water conservation, livelihood diversification, changing livestock and aquaculture practices, changing cropping practices, patterns and planting dates, reliance on social networks, grass-root approaches

(continued)

Table 3.4 (continued)

SETS domain	SETS code	SETS component	Strategies exemplifying component
	S5	Economic	Financial incentives (including taxes and subsidies), insurance (including index-based weather insurance schemes), catastrophe bonds, revolving funds, payments for ecosystem services, water tariffs, savings groups, microfinance, disaster contingency funds, cash transfers
	S6	Legal	Land zoning laws, water regulations and agreements, requirements to support disaster risk reduction, laws to encourage insurance purchasing, defining property rights and land tenure security, eminent domain protected areas, marine protected areas, fishing quotas, patent pools and technology transfer
	S7	Institutional	New research, cross-institutional coordination, partnerships, changes in institutional structure. *S7 captures interactions among agencies (including governmental institutions, non-governmental organizations, and public–private partnerships)
Ecological	E1	Ecosystem-based	Ecological restoration, wetland and floodplain conservation and restoration, increasing biological diversity, afforestation and reforestation, conservation and replanting mangrove forest, bushfire reduction and prescribed fire, assisted migration or managed translocation, ecological corridors, ex situ conservation and seed banks, green and open space
	E2	Green infrastructure	Green infrastructure (e.g., shade trees, green roofs), urban gardens, rain gardens, engineered or constructed ecosystem services
	E3	Ecosystem management practices	Community-based natural resource management, adaptive land use management, controlling overfishing, fisheries co-management, ecosystem focused plan, management of natural resources and ecosystem features/services

(continued)

Table 3.4 (continued)

SETS domain	SETS code	SETS component	Strategies exemplifying component
Technological	T1	Built environment planning and design	Urban planning and design, design storm, building codes, standards, engineering, planning and design codes, certification, and specification
	T2	Engineered infrastructure	Seawalls and coastal protection structures, flood levees, sewage works, improved drainage, beach nourishment, pavement, physical buildings, green infrastructure, solar shade, flood and cyclone shelters, elevate buildings, new system construction and existing system modification and improvement
	T3	Infrastructure operation and maintenance	System inspection and monitoring, operator training program, facility and equipment maintenance/repair, drainage cleaning, best management practices (BMPs)
	T4	Technological solution development and improvement	New crop and animal varieties, genetic techniques, traditional technologies, efficient irrigation, water-saving technologies, conservation agriculture, food storage and preservation facilities, hazard mapping and monitoring technology, early warning systems, building insulation, mechanical and passive cooling, renewable energy technologies, second-generation biofuels

viable adaptation actions. Shaping climate adaptation goals and instigating governance strategies by integrating social, ecological, and technological domains in a systems perspective is essential for building urban resilience, and ultimately, for enabling transformation to sustainable pathways toward the resilient future.

Table 3.5 Example of coded strategies using the SETS codebook. To maintain inter-coder reliability, multiple coders analyzed and reviewed each strategy following the suggested codebook in Table 3.4. Before analysis, selected coders were trained according to standardized coding protocol and the codebook to maintain coding coherency across various documents and among coders. SETS codes correspond to SETS components set out in Table 3.4

City	Extracted strategy	Levers	SETS code	Exogenous drivers
Baltimore	"Restore and protect at least one mile per year of streams and river basins in floodplains and stream valleys" (City of Baltimore 2009, p 139)	Flood infrastructure	E1	Flooding Non-specific
New York City	"NYCHA to execute a resiliency program across 33 public housing developments, which will include the elevation and hardening of building systems, flood-proofing, and upgrading infrastructure" (City of New York 2015, p 231)	Research and plan development; Building design; Flood infrastructure	S7; T1; T2	Flooding Urban
Phoenix	"Implement a water harvesting program through the location, design and construction of dual functioning stormwater retention facilities with enhanced recharge elements designed into the basin…as a part of maintaining a Drought Management Plan in conjunction with SRP & APS to lessen the impact of drought" (Maricopa County 2015, p 402)	Stormwater capture; Groundwater recharge; Intergovernmental coordination	S7; E2; T2	Flooding Urban; Drought

(continued)

Table 3.5 (continued)

City	Extracted strategy	Levers	SETS code	Exogenous drivers
San Juan	Review and modify the public transportation routes depending on the effects of sea-level rise, storm surges and floods [Translated from an original quote in Spanish] (PRCCC 2015, p 40)	Transportation infrastructure	S1; T1	Flooding Coastal

References

Araos M, Berrang-Ford L, Ford JD et al (2016) Climate change adaptation planning in large cities: a systematic global assessment. Environ Sci Policy 66:375–382. https://doi.org/10.1016/j.envsci.2016.06.009

Berbés-Blázquez M, Mitchell CL, Burch SL et al (2017) Understanding climate change and resilience: assessing strengths and opportunities for adaptation in the global South. Climatic Change 141(2):227–241. https://doi.org/10.1007/s10584-017-1897-0

Bulkeley H (2010) Cities and the governing of climate change. Annu Rev Environ Resour 35(1):229–253. https://doi.org/10.1146/annurev-environ-072809-101747

Burch S, Mitchell CL, Berbés-Blázquez M et al (2017) Tipping toward transformation: progress, patterns and potential for climate change adaptation in the Global South. J Extreme Events 04(01):1750003. https://doi.org/10.1142/S2345737617500038

Cámara Chilena de la Construcción (2010) Actualización Plan Regulador Comunal de Valdivia. Delegación de Valdivia. https://biblioteca.cchc.cl/datafiles/22341-2.pdf. Accessed 07 Jul 2020

Carmin J, Anguelovski I, Roberts D (2012) Urban climate adaptation in the global South. J Plann Educ Res 32(1):18–32. https://doi.org/10.1177/0739456X11430951

Carpenter SR, Booth EG, Gillon S et al (2015) Plausible futures of a social-ecological system: Yahara watershed, Wisconsin, USA. Ecol Soc 20(2):10. https://doi.org/10.5751/ES-07433-200210

City of Baltimore (2009) Comprehensive master plan. Maryland Department of Planning, Baltimore. https://www.baltimorecity.gov/sites/default/files/070909_CMPfullplan.pdf. Accessed 04 Jul 2020

City of New York (2015) One New York: The plan for a strong and just city. OneNYC 2050. https://www.nyc.gov/html/onenyc/downloads/pdf/publications/OneNYC.pdf. Accessed 04 Jul 2020

City of Portland (2020) 2035 comprehensive plan. Oregon Bureau of Planning and Sustainability, Portland. https://www.portland.gov/bps/comp-plan/2035-comprehensive-plan-and-supporting-documents#toc-2035-comprehensive-plan-as-amended-through-march-2020-. Accessed 07 Jul 2020

da Silva J, Kernaghan S, Luque A (2012) A systems approach to meeting the challenges of urban climate change. Int J Urban Sustain Dev 4(2):125–145

Denton F, Wilbanks TJ, Abeysinghe AC et al (2014) Climate-resilient pathways: Adaptation, mitigation, and sustainable development. In: Field CB, Barros VR, Dokken DJ et al (eds) Climate change 2014 impacts, adaptation, and vulnerability. Cambridge University Press, Cambridge, pp 1101–1131. https://doi.org/10.1017/CBO9781107415379.025

Feagan M, Matsler M, Meerow S et al (2019) Redesigning knowledge systems for urban resilience. Environ Sci Policy 101:358–363. https://doi.org/10.1016/j.envsci.2019.07.014

Folke C, Hahn T, Olsson P et al (2005) Adaptive governance of social-ecological systems. Annu Rev Environ Resour 30(1):441–473. https://doi.org/10.1146/annurev.energy.30.050504.144511

Galafassi D, Daw TM, Munyi L et al (2017) Learning about social-ecological trade-offs. Ecol Soc 22(1):2. https://doi.org/10.5751/ES-08920-220102

Grimm NB, Cook EM, Hale RL et al (2015) A broader framing of ecosystem services in cities: Benefits and challenges of built, natural or hybrid system function. In: Seto KC, Solecki WD, Griffith CA (eds) The Routledge Handbook of Urbanization and Global Environmental Change. Routledge, New York, pp 203–212. https://doi.org/10.4324/9781315849256

ICLEI Mexico (2015). PACMUN—plan de Acción Climática Municipal

Iwaniec DM, Cook EM, Davidson MJ, Berbés-Blázquez M et al (2020) Integrating existing climate adaptation planning into future visions: a strategic scenario for the central Arizona-Phoenix region. Landscape Urban Plan 200:103820. https://doi.org/10.1016/j.landurbplan.2020.103820

Kim Y, Chester MV, Eisenberg DA et al (2019) The infrastructure trolley problem: positioning safe-to-fail infrastructure for climate change adaptation. Earth's Future 7(7):704–717. https://doi.org/10.1029/2019EF001208

Lempert RJ, Popper SW, Bankes SC (2003) Shaping the next one hundred years: new methods for quantitative, long-term policy analysis. RAND Corporation. https://www.rand.org/pubs/monograph_reports/MR1626.html. Accessed 03 Jun 2020

Maricopa County (2015) Multi-Jurisdictional Hazard mitigation plan. Maricopa County Department of Emergency Management. https://www.maricopa.gov/DocumentCenter/View/5118/Hazard-Mitigation-Plan-PDF. Accessed 04 Jul 2020

Markolf SA, Chester MV, Eisenberg DA et al (2018) Interdependent infrastructure as linked social, ecological, and technological systems (SETSs) to address lock-in and enhance resilience. Earth's Future 6(12):1638–1659. https://doi.org/10.1029/2018EF000926

McPhearson T, Haase D, Kabisch N et al (2016) Advancing understanding of the complex nature of urban systems. Ecol Indic 70:566–573. https://doi.org/10.1016/j.ecolind.2016.03.054

Muñoz-Erickson T, Miller C, Miller T (2017) How cities think: knowledge co-production for urban sustainability and resilience. Forests 8(6):203. https://doi.org/10.3390/f8060203

Puerto Rico Climate Change Council (PRCCC) (2015) Ruta hacia la Resiliencia: Guía de Estrategias para la Adaptación a los Cambios Climáticos. In Díaz EL, Jacobs KR, Marrero VI (eds) Programa de Manejo de la Zona Costanera. https://pr-ccc.org/download/Ruta-hacia-la-resiliencia-webview.pdf. Accessed 04 Jul 2020

Reckien D, Salvia M, Heidrich O et al (2018) How are cities planning to respond to climate change? Assessment of local climate plans from 885 cities in the EU-28. J Clean Prod 191:207–219. https://doi.org/10.1016/j.jclepro.2018.03.220

Southeast Florida Regional Climate Change Compact (2015) Regional climate action plan. Broward, Miami-Dade, Monroe, and Palm Beach Counties. https://southeastfloridaclimatecompact.org//wp-content/uploads/2014/09/regional-climate-action-plan-final-ada-compliant.pdf. Accessed 04 Jul 2020

United Nations Office for Disaster Risk Reduction (2017) How to make cities more resilient: a handbook for local government leaders. Switzerland, Geneva

Wamsler C, Luederitz C, Brink E (2014) Local levers for change: Mainstreaming ecosystem-based adaptation into municipal planning to foster sustainability transitions. Global Environ Change 29:189–201. https://doi.org/10.1016/j.gloenvcha.2014.09.008

Wiek A, Iwaniec D (2014) Quality criteria for visions and visioning in sustainability science. Sustain Sci 9(4):497–512. https://doi.org/10.1007/s11625-013-0208-6

Yeowon Kim is an interdisciplinary scientist whose work explores infrastructure systems and institutional governance for sustainable urban futures influenced by resilience theory, civil and environmental engineering, and systems science. She is a postdoctoral research fellow of the Urban Systems Lab at The New School and an adjunct faculty in the School of Sustainability at

Arizona State University. Her research interests include advancing resilient infrastructure development facing extreme weather events and understanding how social, ecological, and technological systems are interacting with one another to adapt to changing urban environment. Her work spans integrating quantitative modeling, participatory engagement of stakeholders, and institutional knowledge systems analyses in climate adaptation efforts.

Lelani M. Mannetti is a Postdoctoral Research Associate at the Urban Studies Institute, conducting comparative research across nine US and Latin American cities as part of the Urban Resilience to Extreme Events Sustainability Research Network. Alongside city stakeholders, she helps promote visionary thinking through the co-production of desirable and plausible future scenarios focused on resilience, sustainability, and equity. Dr. Mannetti is also a Fellow on the Intergovernmental Science-Policy Plat-form on Biodiversity and Ecosystem Services Values Assessment, a methodological assessment of the diverse conceptualization of multiple values of nature and its benefits.

David M. Iwaniec is an Assistant Professor of Urban Sustainability at the Urban Studies Institute, Andrew Young School of Policy Studies at Georgia State University. He is a sustainability scientist researching anticipatory and systems approaches to advance urban sustainability, resilience, and justice. His work focuses on the co-development of scenarios and transition pathways for positive futures of urban transformation

Nancy B. Grimm studies the interactions between climate variation and change, human activities, and ecosystems. Her interdisciplinary urban and stream ecosystem research focuses on disturbance, resilience, and bio-geochemical processes. Dr. Grimm is Regents Professor at Arizona State University and currently co-directs the Urban Resilience to Extremes Sus-tainability Research Network (UREx SRN). She was president (2005-06) and is a fellow of the Ecological Society of America, as well as a fellow of the American Geophysical Union (AGU), the American Association for the Advancement of Science (AAAS), and the Society for Freshwater Science. She is an elected member (2019) of the US National Academy of Sciences.

Marta Berbés-Blázquez is an Assistant Professor at the School for the Future of Innovation at Arizona State University. Her research considers the human dimensions of social–ecological transformations in rural and urban ecosystems with an emphasis on vulnerable populations. Her work is informed by resilience thinking and political ecology at a conceptual level, and it is practically oriented toward qualitative, participatory, and anticipatory research methods. Specific topics of expertise include power dynamics and access in ecosystem services, scenario planning, resource extraction, eco-health, climate change adaptation, and transformation.

Sam Markolf is an Assistant Research Professor within the School for Sustainable Engineering and the Built Environment at Arizona State University and a Research Fellow within the NSF sponsored Urban Resilience to Extremes Sustainability Research Network (UREx SRN). His research broadly focuses on applying systems-level analysis to sustainability and resilience challenges facing cities and infrastructure systems. Current pro-jects include modeling the impacts of, and responses to, various disrup-tion scenarios for inter-city transportation systems, as well as analyzing the extent to which interconnected social-ecological-technological sys-tems (SETS) can enhance (or hinder) the resilience of cities and infrastruc-ture systems to extreme events. He earned a B.S. in Chemical Engineering from the University of Texas at Austin, M.S. in Civil & Environmental Engi-neering from Carnegie Mellon University, and a joint-Ph.D. in Civil & Envi-ronmental Engineering and Engineering & Public Policy from Carnegie Mellon University.

Chapter 4
Mapping Vulnerability to Weather Extremes: Heat and Flood Assessment Approaches

Zoé A. Hamstead and Jason Sauer

Abstract Assessing present social and biophysical conditions of communities that are at risk of injury due to extreme weather events is an important component of creating future visions of resilience. Spatial patterns of vulnerability to extreme events are manifestations of structural injustice that leave their mark on the built environment and in socio-spatial segregation patterns. Socio-spatial inequity often arises from development practices that favor particular racial and ethnic social groups over others. These segregation patterns are aligned with patterns of exposure to pollution, extreme weather events, and other types of environmental hazards. Spatial vulnerability assessments can be powerful tools for prioritizing where and how cities should make investments for mitigating the impacts of extreme events, and can provide an entry point for asking more fundamental questions about the processes that produce patterns of climate inequity, as well as how to avoid reproducing such processes in the future. Maps express uneven distributions of risk and manifestations of structural inequality in social–ecological–technological systems (SETS). They enable communities to visualize distributional injustice, consider ways in distributions that may be misaligned with cultural values, and develop adaptive practices toward climate justice. Here, we demonstrate approaches for assessing vulnerability to extreme flooding and heat, and show how vulnerability distributions are embedded in landscape patterns that produce uneven risk.

Keywords Climate vulnerability · Vulnerability mapping · Social vulnerability index · Mapping injustice

---

Z. A. Hamstead (✉)
Department of Urban and Regional Planning, School of Architecture and Planning, University at Buffalo, State University of New York, Buffalo, New York, NY, USA
e-mail: zoehamst@buffalo.edu

J. Sauer
School of Life Sciences, Arizona State University, Tempe, AZ, USA

Z. A. Hamstead et al. (eds.), *Resilient Urban Futures*, The Urban Book Series,
https://doi.org/10.1007/978-3-030-63131-4_4

4.1 Vulnerability Frameworks and Spatial Vulnerability Assessments for Resilience

Vulnerability mapping is a way of conveying a particular configuration of biophysical and social conditions that bear relation to a particular form of risk. Vulnerability can be understood in opposition to resilience, but the two concepts tend to emphasize different aspects of SETS (see Chap. 3). Resilience is a term with foundations in mathematics, physics (Norris et al. 2008), and the natural sciences, particularly ecology. It generally means a system's capacity to undergo change while continuing to persist and function via adaptation (Berkes et al. 2003). In the context of disasters, it means the ability to withstand, cope, and recover from an event (Brown 2012).

Vulnerability, on the other hand, is a term more commonly used in the social sciences and as a way of articulating how best to prioritize resources deployed in hazards management practice (Cutter 1996; Morrow 1999). It can have a broad range of meanings, generally referring to three main dimensions: exposure, sensitivity, and adaptive capacity (Adger 2006). The Intergovernmental Panel on Climate Change (IPCC) defines exposure as "[t]he presence of people; livelihoods; environmental services and resources; infrastructure; or economic, social or cultural assets in places that could be adversely affected" (IPCC 2012 p. 559). In addition to what is exposed, the term may also refer to an event (disaster) of some intensity and duration as well as the biophysical conditions that attenuate or exacerbate that event. Sensitivities are conditions that predispose people to risk or enable them to cope with stress; these are typically described in terms of demographic risk factors. The IPCC's definition of coping capacity most closely resembles the concept of sensitivities (by way of its opposite): "[t]he ability of people, organizations and systems, using available skills, resources, and opportunities, to address, manage and overcome adverse conditions" (IPCC 2012 p. 558). Adaptive capacity is "[t]he combination of the strengths, attributes, and resources available to an individual, community, society or organization that can be used to prepare for and undertake actions to reduce adverse impacts, moderate harm, or exploit beneficial opportunities" (p. 556). Sensitivity, coping capacity, and adaptive capacity are closely related and often conflated with one another other. In practice, sensitivities and coping capacities are generally mapped or statistically presented as static conditions, whereas adaptation may involve short, medium, or long-term processes of reconfiguring those conditions based on abilities that are not limited to coping capacities. Vulnerability mapping most often involves overlaying exposure with sensitivity. It brings socio-spatial inequity to light, as differential vulnerabilities are often associated with development patterns that lead to spatial segregation of racial and other social groups.

Although differential social and geographic risks are manifestations of processes that play out over time, the vulnerability dimensions are often treated as temporally static (though geographically dynamic). Yet, conditions of vulnerability come about as a result of social, economic, and political processes, and those processes themselves can change as a result of extreme events in ways that create differential vulnerabilities to hazards (Wisner et al. 2003). In contrast with the vulnerability framework, the temporal dimension is conceptually central to the resilience framework. Yet, its application to disasters practice tends to treat a set of system properties rather than processes (Brown 2012) and to oversimplify social dynamics that produce risk. Our approach to vulnerability mapping draws largely from the social science scholarship. However, in our view, the purpose of producing and analyzing static maps that convey patterns of vulnerability is to provide not only an entry point for spatially prioritizing interventions but also for investigating the development, policy, exclusionary, and biophysical processes that produce patterns of vulnerability. We consider both efforts essential for enhancing resilience by building adaptive capacities that address the needs of people most at risk of injuries triggered by extreme events.

4.1.1 Extreme Heat Vulnerability

Vulnerability to extreme heat is characterized by local climate variability and land cover patterns (biophysical exposure); social constraints (neighborhood socioeconomic sensitivity); and individual and household-level social capital, knowledge, and practices (capacity to change behaviors and conditions in response to heat threats) (Wilhelmi and Hayden 2010). Owing to prevailing design decisions, cities absorb a majority of incoming solar radiation. Compared with less urban areas, cities have less capacity to moderate temperatures via ecological processes. Together, these features contribute to a phenomenon known as the urban heat island effect, or the phenomenon that cities are detectably hotter than their surrounding peri-urban and rural areas. That is, if one were to look at a temperature map of a region, the urban areas would appear as islands of hotter temperatures in a sea of cooler, more rural areas. However, research into the heat island phenomenon emphasizes that even within a given urban heat island there is substantial spatial heterogeneity. The differences in temperature between these micro-urban heat islands and the coolest areas of a city may be as great as those of the broader urban heat island and its surrounding rural region. Accordingly, this spatial heterogeneity of temperature produces differential effects. Since thermal properties (e.g., absorption, storage, radiation) are closely linked to the composition of built and natural materials, the presence of such materials can be used to understand variation in heat exposure and the extent to which landscapes contribute to such exposures (Hamstead et al. 2016). People who live in neighborhoods replete with asphalt parking lots and little tree canopy are exposed to higher temperatures and are consequently at disproportionate risk of heat-related illness.

At the same time, the characteristics of the urban social system produce sensitivity and adaptive capacity, and these system parameters interact. Klinenberg's (2002) study of the heat wave that struck Chicago in 1995 found that people living in social isolation tended to be most affected, though a more full account of why over 700 people died during that heat event and why some neighborhoods were more affected than others had to do with broader economic and policy forces, including economic cycles of community abandonment and lack of disaster management response. Other factors linked to heat-related mortality, illness, and distress calls at the individual and neighborhood levels that include surface temperatures, impervious land cover, green space, minority race and ethnicity, linguistic isolation, age, level of educational attainment, income, disability, housing conditions, housing values, vacant households, and rates of access to air conditioning in the home (Hattis et al. 2012; Madrigano et al. 2015; Rosenthal et al. 2014; Smargiassi et al. 2009; Uejio et al. 2011).

4.1.2 Flood Vulnerability

Vulnerability to extreme flooding is characterized by local climate variability, topography, and drainage infrastructure (biophysical exposure); social constraints (neighborhood socio-economic sensitivity); and individual and household-level social capital, knowledge, and practices (capacity to change behaviors and conditions in response to flooding threats). Delineating biophysical exposure to urban flooding is complex in both theory and practice. It involves interactions between regional climate and weather, as well as local topographic and drainage system characteristics. There are numerous types of flood exposure that may result from common or unique sources. Coastal flooding is due to tidal or storm surges bringing water into the city; fluvial flooding is due to overtopping riverbanks or levees, driven by rainfall across the relevant watersheds; pluvial flooding is due to rainfall intensity exceeding the performance capacity of the city's stormwater management system, or from evading this system entirely or in part. In many cities across the United States and the globe, it is generally expected that cities develop drainage and levee infrastructure to prevent or manage exposure to flooding from all sources. This expectation has been met primarily through technological measures such as pipes, canals, and hardened shorelines.

However, cities operate with limited information and resources, and in highly dynamic circumstances. Regional precipitation patterns change, sea levels rise, impermeable surfaces proliferate, and drainage infrastructure ages or proves to be maladapted to evolving urban configurations; cities must contend with several or all of these dynamic exacerbators of flood exposure. Further, a range of governmental and non-governmental actors engage in flood mitigation, from individuals physically rerouting pluvial flood water from homes and businesses by placing sand bags in front of building entrances, to neighborhoods ensuring adequate drainage system performance by coordinating pre-storm trash and debris removal efforts.

Differential social and demographic factors can cause differential impacts of extreme flooding on populations, even given the same forms and degrees of exposure. Home renters, for example, may have little to no agency to make modifications to the structure of their homes, relying instead on the diligence of the landlords who may not adapt their properties to address tenant flooding concerns (Morrow 1999). At the same time, the potential agency of landlords may be stymied by factors such as low income from rents, an unpaid mortgage on the building, advanced age, and disability (Cutter et al. 2003).

4.2 Role of Vulnerability Maps

Spatial vulnerability assessments are a way of conveying distributional patterns of exposure, sensitivity, and (to a lesser extent) adaptive capacity along the spatial dimension. They emphasize the overlay of social, ecological, and technological contexts of that vulnerability. The process of mapping vulnerability involves identifying appropriate indicators of the hazard, people, infrastructure, and ecosystems that are exposed to that hazard, varying levels of sensitivity and adaptive capacity, and aggregating all of this information. The ability to map these indicators depends on spatially-explicit data, which may not be available for all cities or for all areas of a single city. Sensitivity and adaptive capacity indicators are place-based. What constitutes as low income or racial minority, and as governance processes or emergency response protocols can vary widely in cities across the globe. Thus, the particular indicators used in vulnerability mapping are only transferable to a limited extent from one city to the next.

Ideally, cities would use current vulnerability maps as a baseline from which to plan for managing vulnerabilities over the long term, integrating results from regional climate models, a comprehensive suite of exposure models, and projections of future demographics. For instance, climate and weather models that forecast intensities and durations of storms could be used to produce estimates of runoff generated by various urban surfaces (such as through the rational runoff method), the locations and depths of flooding in the city's land surface depressions (Balstrøm and Crawford 2018), and measurements of the flooding that occurs along and outward from the stormwater drainage system through which stormwater is routed (such as via the United States Environmental Protection Agency's Storm Water Management Model, or SWMM). Demographic changes—such as an aging population—are also important for projecting relationships between flood exposure and capacities for coping with flooding. Spatial planning for managing these vulnerabilities includes climate and human population dynamics.

In practice, cities may have only coarse spatial resolution of land cover and temperature data, no estimates of how the regional climate is changing, exposure estimations from only one hazard model, no estimation of the future drainage network or land cover, and no spatially explicit projections of demographic changes. Even if a city does have a model for estimating current flooding exposure, the drainage system likely does not perform according to its design standards, due to structural deterioration or fouling of waterways by debris. Such limitations as these leave cities unable to give a full accounting of their current biophysical exposure and social vulnerability, let alone that of their future forms. Thus, cities tend to plan according to their current configuration, or for a similar configuration in the short term. For instance, New York City prioritized street tree planting in communities that rank high on an of-the-moment social vulnerability index (SVI) for heat (City of New York 2017).

In addition to providing a baseline from which to engage in future spatial planning, vulnerability assessment may also simply raise the visibility of disproportionate burden of extreme events and the socio-spatial distributions of risk (Walker 2009) in ways that—along with direct information about people's experiences—provide an evidentiary basis for community conversations about environmental justice. Moreover, mapping can be part of an inductive approach whereby patterns of socio-spatial segregation can be used to generate hypotheses about the economic, exclusionary, and policy processes that are producing and reproducing uneven risk in cities.

4.3 Urban Resilience to Extremes (UREx) Assessments and Mapping Methodologies

4.3.1 Vulnerability Assessments

As mentioned above, the selection of exposure and sensitivity variables is an imperfect process, and one which generally relies on published literature or expert opinion. Ideally, sensitivity indicators are selected on the basis of studies showing which risk factors are most closely related to an undesirable outcome, or on input from residents who experience such outcomes. For instance, many studies identify risk factors of heat-related mortality, illness, and distress calls (Bell et al. 2008; Hattis et al. 2012; Hondula et al. 2015; Kovats and Hajat 2008; Madrigano et al. 2015; Medina-Ramón et al. 2006; Rosenthal et al. 2014; Smargiassi et al. 2009; Uejio et al. 2011) and high financial burden due to flood-related property destruction (Balica et al. 2012). However, indicators of risk are highly contextual. In most U.S. cities, heat-related fatality rates are higher in low-income communities of color, but these constructs are not necessarily meaningful in all U.S. communities, let alone in Global South contexts. Vulnerability assessments that lack scientific or community experience-based informational resources from which to draw may be conducted on the basis of geographically-proximate studies or more general consensus about what factors are related to the production of environmental risk. Once indicators of vulnerability are constructed, they can be aggregated into a single vulnerability index (e.g., Johnson et al. 2012), and spatial clusters of vulnerability can be identified (e.g., Hamstead et al. 2018; Inostroza et al. 2016). Spatial scales at which indices are assessed include the census tract (Chow et al. 2012; Reid et al. 2009; Rinner et al. 2010), census block group (Bradford et al. 2015; Johnson et al. 2012; Uejio et al. 2011), Canadian dissemination area (e.g., in Canada; Rinner et al. 2010), or an even finer block scale where available (e.g., in Santiago, Chile; Inostroza et al. 2016). Here, we describe an application of heat and flooding vulnerability assessments in Hermosillo, Mexico.

4.3.1.1 Hermosillo Heat Vulnerability Assessment

To assess heat vulnerability in Hermosillo, Mexico we extracted the Landsat 8 thermal band for September 19, 2013 in order to produce a surface temperature image, and used the Senora 2010 census data at the block-level to derive an estimate of populations exposed to hot microclimates. The temperature and total population were transformed to indices on a 0 to 1 scale and aggregated into an index of exposure.

To more precisely map locations of people who are sensitive to extreme heat, we first created population maps at a finer scale than the census block-level data available for Hermosillo. Block groups and other enumeration units include all residential and non-residential areas, and rarely reflect actual population distributions (Sleeter and Gould 2007). Dasymetric mapping is an interpolation technique that disaggregates population data by empirically sampling population values in an ancillary dataset (typically of land use) which represents the population statistical surface at a finer scale than that of the original population data. Based on this sampling procedure, weights are assigned to the classes of the ancillary dataset, and population values are disaggregated from the original spatial resolution to the finer resolution according to these derived weights (Mennis 2003). Since geographic units of analysis are often arbitrarily defined in relation to their applications and analyses, this approach is particularly useful for addressing ways in which the modifiable areal unit problem (MAUP) can mask problems of environmental justice (Mennis 2002).

The following variables comprised the sensitivity index: population <5 years of age, population >65 years of age, houses without electricity, houses without tap water, houses with at least one vehicle, illiterate adult population over 15 years, population without health services, unemployed population, disabled population, and total population. These variables were disaggregated using a land use parcel dataset which indicates high, medium, and low-density residential parcels, as well as mixed use and non-residential. Using the indexing procedure described for exposure above, we created indices for disability, quality of life (no tap water, no electric, age index, education/literacy index, health service access, vehicle access index), and economic constraint. These were then normalized and combined to an overall sensitivity index, which was then combined with a normalized exposure index to generate an index of overall vulnerability (Fig. 4.1).

4.3.1.2 Hermosillo Flood Vulnerability Assessment

For flood vulnerability, we used the city's high-resolution topographic information (1 m resolution LiDAR) and estimates from the localized constructed analog (LOCA) model to develop a blue spot model for flooding (Balstrøm and Crawford 2018). This blue spot model combines a digital elevation model (DEM) input and the sum of precipitation that occurs over the course of a rainfall event of some return period of interest (in our case, a 100-year return period storm), and the results are a shapefile of the locations of depressions in the landscape that fill to capacity, and tabular information about the depths of these filled depressions. We produced a final relative exposure map by calculating the percentage of a given census block covered by areas that experience blue spot flooding (Fig. 4.2).

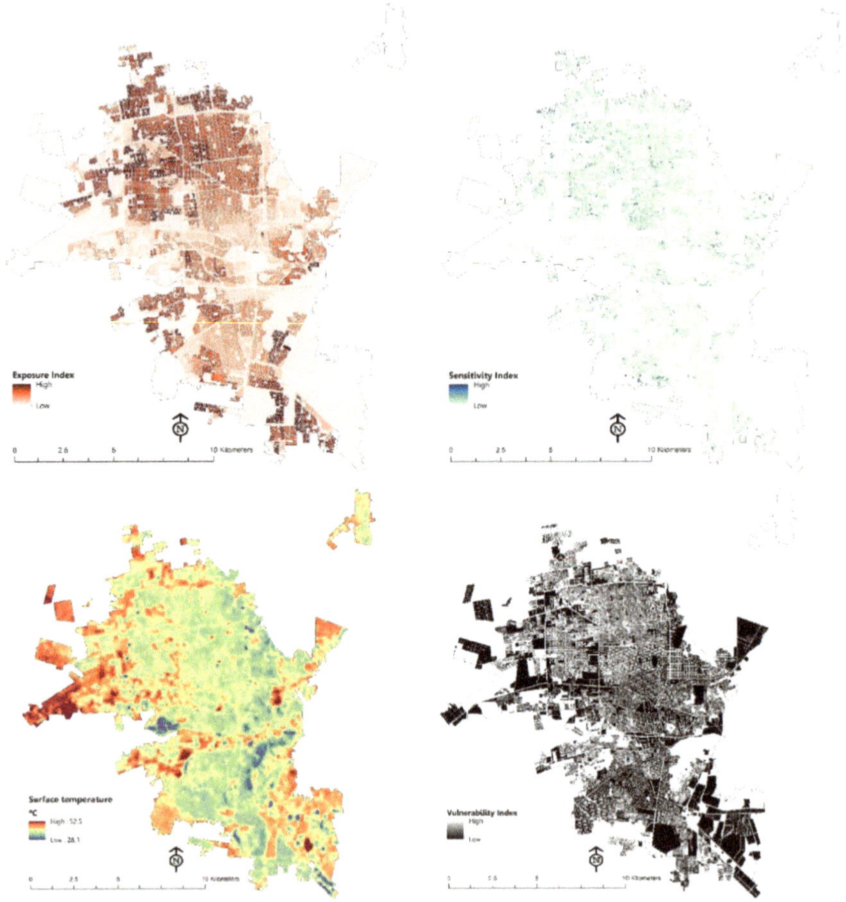

Fig. 4.1 From upper right to lower left: Exposure index, sensitivity index, surface temperature, and overall vulnerability index

Additionally, we considered technological sources of flooding in the city, specifically those which might be caused by failure of the drainage infrastructure. Engineers who worked with us relayed that older, narrower pipe elements of the drainage network would be more likely to fail or be overwhelmed than newer, wider ones. Thus, using a polyline file provided by the city that included the pipe diameter and age, we determined the relative exposure of the city to flooding due to differences in the age and dimensions of pipes in its drainage network (Fig. 4.3).

For sensitivity indicators, we used an inductive approach and polled local flooding experts on the social and demographic factors they saw as critical for determining differences in sensitivity and adaptability among populations (Table 4.1).

We used a 2010 census data shape file that aggregated these factors at the census block level. We assigned factors equal weight, normalized them between 0 and 1

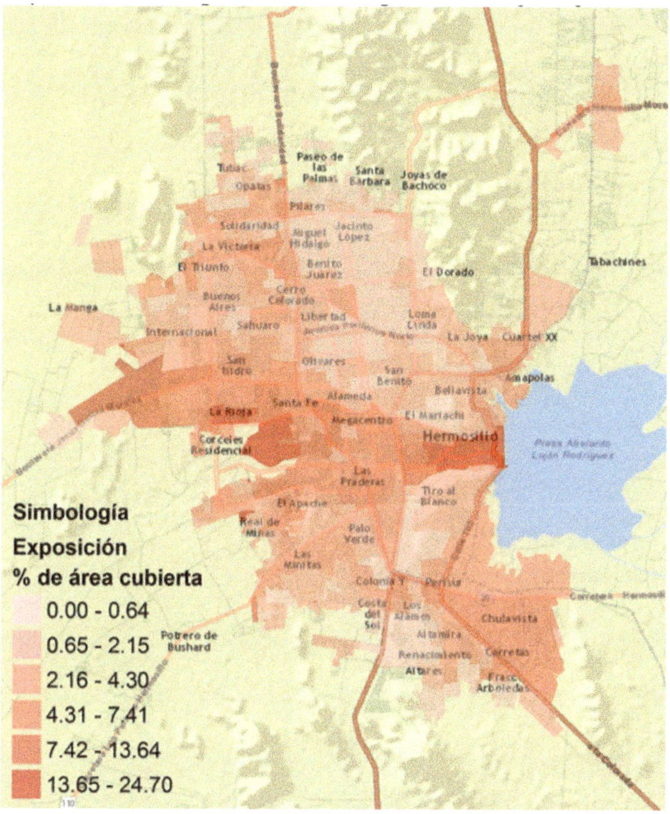

Fig. 4.2 Relative flood exposure map indicating the percent area of each census block covered by estimate flood areas according to the blue spot model

(where 0 was the lowest value of a given factor across the entire city and 1 was the highest value of a given factor across the entire city), and aggregated them. Thus, census blocks with greater totals had greater relative vulnerability than census blocks with lower totals (Fig. 4.4).

These blue spot exposure, technological vulnerability, and social vulnerability indicators were combined such that:

$$\begin{aligned}\text{Combined vulnerability} &= \text{blue spot exposure} \\ &\quad * (\text{technological vulnerability} + \text{social vulnerability})\end{aligned}$$

This combined vulnerability was then normalized between 0 and 1, such that 0 was the lowest overall vulnerability value among all the census blocks, and 1 was the greatest overall vulnerability among all the census blocks (Fig. 4.5).

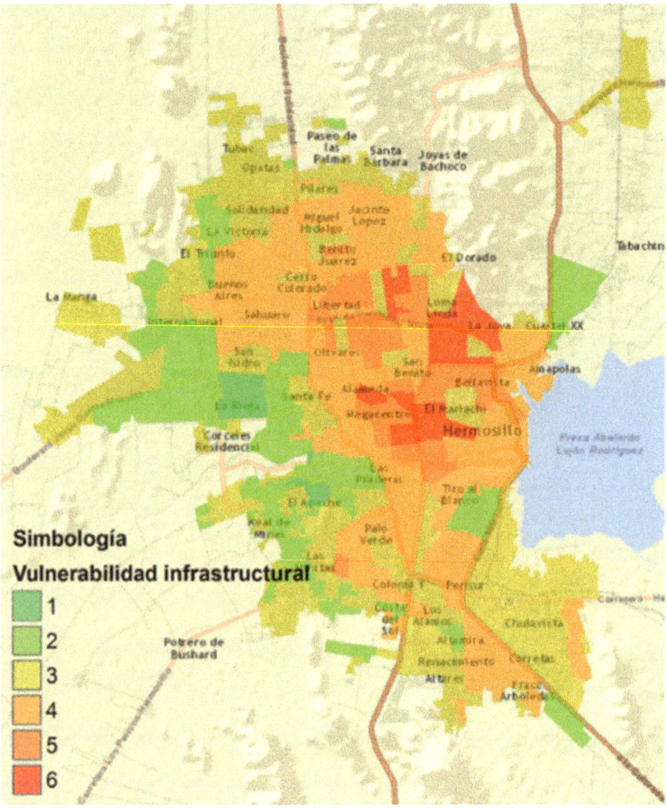

Fig. 4.3 Relative infrastructural vulnerability map indicating areas more (red) and less (green) at risk of being overwhelmed or failing due to extreme rainfall

Table 4.1 Types of social vulnerability indicators used in assessing overall social vulnerability of the city, as well as the forms of vulnerability (sensitivity or adaptability) they represent

Selected social vulnerability indicator	Type(s) of vulnerability
Percent population illiterate	Adaptability
Percent population with some different capacity	Sensitivity/Adaptability
Percent infants (1 to 12 months)	Sensitivity
Percent older than 65	Sensitivity
Household density	Sensitivity
Impoverished ("marginación social")	Sensitivity/Adaptability

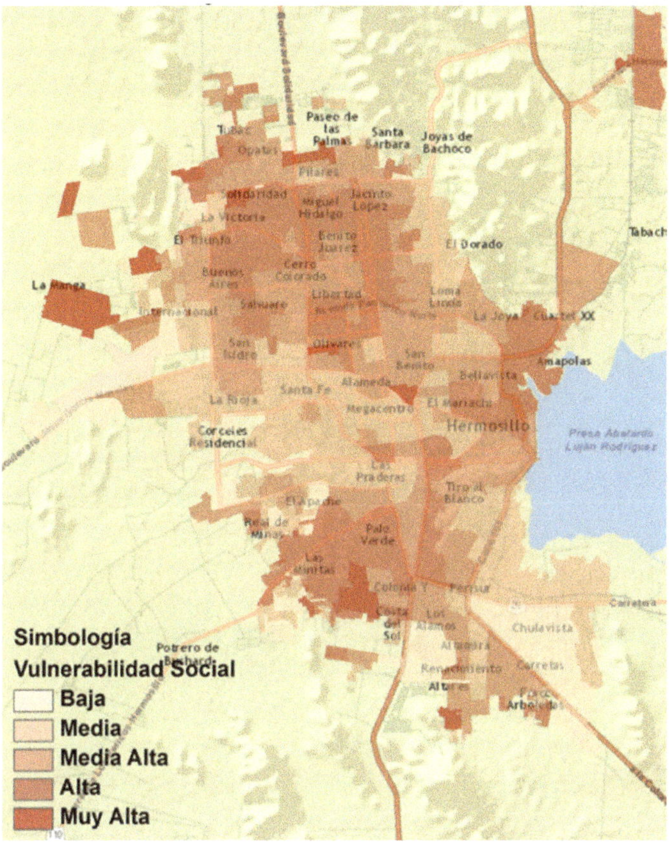

Fig. 4.4 Map showing relative social vulnerability at the census block level. "Baja" indicates low social vulnerability; "Muy Alta" indicates very high social vulnerability

4.3.2 Mapping Urban Landscapes

To inform potential landscape-based interventions in vulnerable locations, we used a landscape-based heat exposure indicator based on the Structure of Urban Landscapes (STURLA) classification (Hamstead et al. 2016) in UREx cities. (For more information about how STURLA is used in future heat projections, see Chap. 9.) STURLA comprised landscape composition elements—including built and natural components—that are common in a given urban environment. The approach involves constructing landscape classes comprising tree canopy, grass/shrub, water, bare soil, paved, low-rise buildings (1–3 stories), mid-rise buildings (4–9 stories), and high-rise buildings (>9 stories). The most abundant classes are defined as those which comprise a majority of land area. For example, in Portland, Oregon, USA, 12 classes account for 90% of the city's land area, revealing a range of temperature signatures from 21.4 to 33.4 °C (Figs. 4.6 and 4.7).

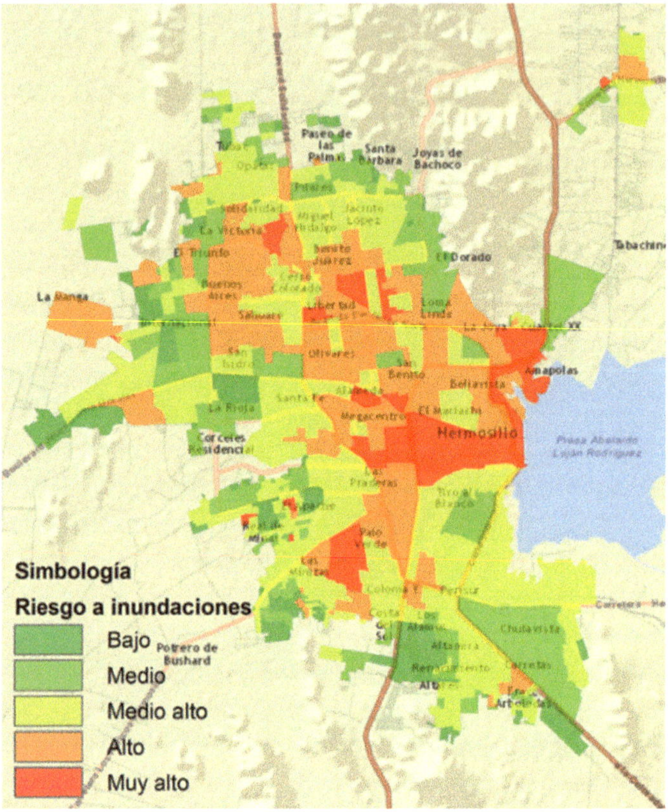

Fig. 4.5 Relative combined vulnerability to flooding in Hermosillo at the census block level. "Bajo" indicates low combined vulnerability; "Muy alto" indicates very high combined vulnerability

Landscape mapping can be used to build on vulnerability assessments by indicating locations where modifiable components of the built environment can be transformed to better support heat mitigation. For instance, areas where pavement predominates could be landscaped with trees, vegetation, and water features, and high-rise buildings could be developed for shade and painted with white roofs in order to reduce albedo. This mapping technique helps to link spatial vulnerability with the built environment features that partially account for it, and inform spatial planning to mitigate micro-urban heat islands.

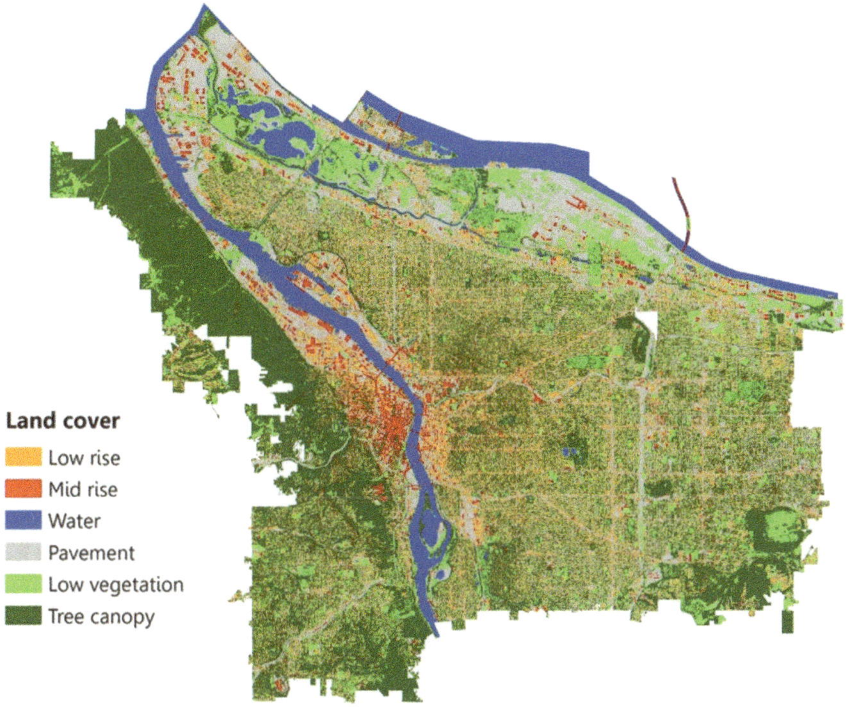

Fig. 4.6 Spatial distribution of land covers in Portland, Oregon, USA

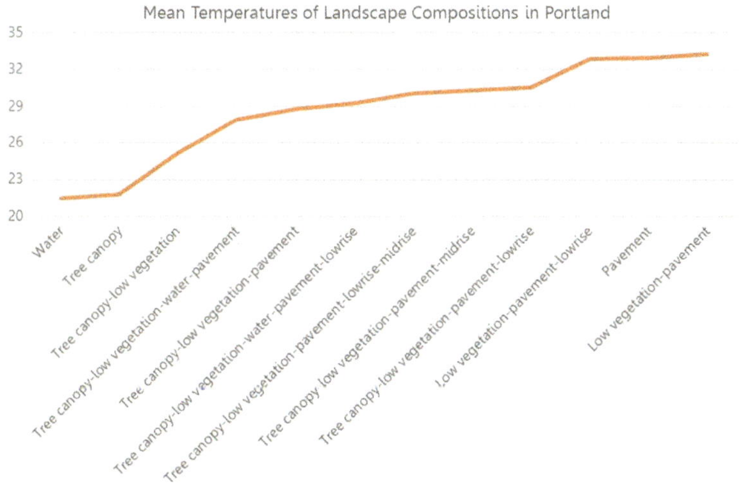

Fig. 4.7 Distribution of temperature across landscape classes in Portland, Oregon, USA. Average temperature (°C) of the frequently occurring landscape compositions, or adjacent land cover combinations that occur across the landscape

4.3.3 Mapping Extreme Event Injustice

We analyzed which populations are disproportionately exposed to hot micro-climates in New York City by performing cluster analysis, which identifies significantly clustered features for which the difference between neighborhood-level values and the sum of all values is too large to be the result of chance. To be a statistically significant hot spot, the feature must have a high value (e.g., high temperature value) and be surrounded by other features with high values. Alternatively, cold spots emerge where features with low values are surrounded by other features with low values. Using this approach, we find that African Americans and Hispanics are more likely to live in hot micro-climate clusters compared with the population as a whole (Figs. 4.8 and 4.9).

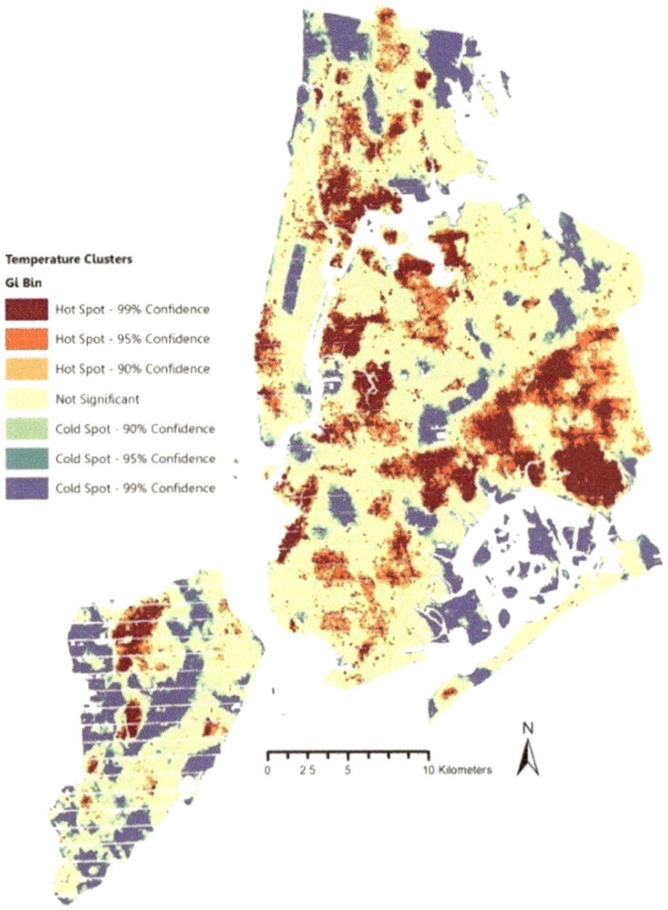

Fig. 4.8 Hot and cold temperature clusters in New York City. Confidence levels indicate the probability that a spatial cluster does not occur due to chance

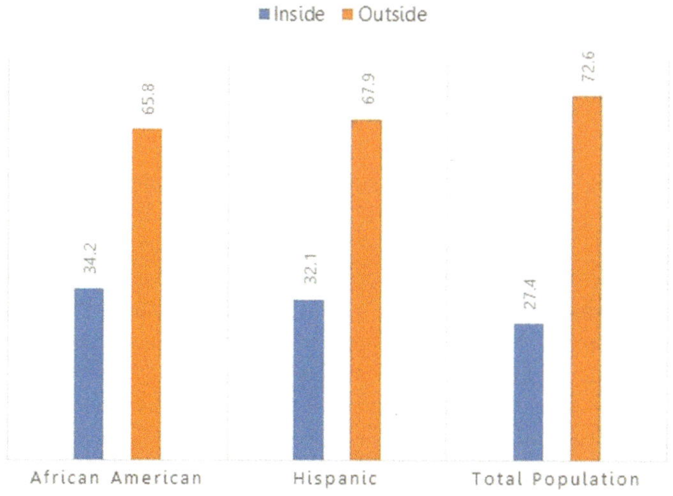

Fig. 4.9 Populations of minority race and ethnicity who live within clusters of hot micro-climates, compared with the population as a whole

Mapping extreme event injustice helps to characterize the extent to which urban development policy has disfavored racial and ethnic groups through both formal and informal policies. For instance, throughout the twentieth century in the USA, racial zoning and mortgage lending practices were used to concentrate African Americans in communities with industrial land uses, and large, heavily-trafficked highways were built through African American communities. These exclusionary housing practices and segregation tactics partially contributed to a landscape of uneven heat burden and likely the discrepancy in heat-related deaths between whites and blacks in the USA (Berko et al. 2014). While vulnerability assessments highlight areas where multiple risk factors overlap, mapping injustice helps to identify which social groups should be most critically engaged to curb the health and economic impacts of extreme events.

4.4 Conclusion

Assessing distributions of risk to extreme weather events in the context of SETS is an important step toward creating future visions of resilience. Maps express uneven distributions of risk and ways in which structural inequality manifests itself in SETS. Maps enable communities to assess ways in which these manifestations misalign with cultural values, and to develop adaptive practices that better represent these values. However, not all values can be expressed spatially, and thus it is important to combine mapping activities with storytelling and other approaches for articulating values. Moreover, vulnerability mapping often focuses on current rather than future conditions and must be combined with future visioning activities in order to help construct resilient urban futures.

In addition to these inherent limitations of vulnerability mapping, there are also technical difficulties. Increasingly, cities are developing more comprehensive hydrological models that incorporate more potential sources for urban flooding than have been included in past models. FLO-2D, for example, is a hydrological modeling software that allows the user to wrangle surface runoff generation, tidal inputs, and drainage network flow in a way that previously would have required the creative use of several separate hydrological models. However, in spite of the availability of such software, comprehensive flood models have not been developed for most cities due to data limitations or scarcity, the lack of personnel with appropriate technical expertise, and the cost of software. Even when sophisticated models are feasible, they are nonetheless limited by their data, which always has some degree of spatial imprecision and often does not reflect current landscape conditions. Additionally, our interview work in Hermosillo, Mexico made clear to us that models, even when accurate, may provide a very limited basis on which managers can design measures to reduce flood exposure and impact. Modelers and professionals who work with these models should bear in mind the aphorism that "all models are wrong, but some are useful," (Box 1976) and further that models should be coupled with, rather than displace, on-the-ground and qualitative methods.

An understanding of contemporary vulnerability conditions requires not only a description of the spatial patterns of that vulnerability, but also of the processes which produced it. In cities where socio-spatial segregation is coupled with exposure to extreme weather, there is a need to identify the economic, political, and exclusionary processes that are producing and reproducing uneven risk. Moreover, while scholarship and practice focus on reducing flood and heat vulnerability by limiting or eliminating exposure, evidence suggests that policy actors should focus as much or more attention on reducing vulnerability through the reduction of poverty. In many cities, the impoverished are more likely to live in areas that are exposed to flooding (Winsemius et al. 2018; Mahanta and Das 2017), and impoverished people—who often experience multiple forms of exclusion—are overall more sensitive and maladapted to flooding (Cutter et al. 2000). Moreover, exposure to flooding causes people to become impoverished and deepens the poverty of people who are already impoverished (Mahanta and Das 2017; Carter et al. 2007). Thus, flooding not only disproportionately affects people who are already sensitive to risk but also increases the portion of people who will be sensitive and maladapted to the next flooding event. Intentionally or not, poverty is often presented by vulnerability researchers as a vulnerability metric of equal consideration to other risk factors, which can be optimally reduced in a remotely determined and technocratic way. In contrast, researchers in the fields of political ecology and critical geography have distinguished poverty as a human rights issue, and emphasized that it must be addressed in ways that contend with place-based conceptions of justice (Ajibade and McBean 2014). The rightful naming of poverty as a human rights issue holds cities, states, and nations accountable for wrongs committed, and has opened successfully pursued pathways to poverty reduction (Bryant 2008).

In addition to coupling vulnerability mapping with practices that enable communities to address the structural inequality that pervades many cities, as well as to orient people toward the future, mapping practices could also be integrated with asset-based community development processes by identifying characteristics of resilience. Conveying the biophysical and institutional assets that help to protect people against extreme weather is important both for providing models of best practices and developing a positive sense of place to serve as a foundation for positive futures.

Acknowledgements This work was supported by the Urban Resilience to Extreme Weather-Related Events Sustainability Research Network (UREx SRN; NSF grant no. SES 1444755) and Environmental Protection Agency Science to Achieve Results Fellowship No. 91760801-0. The authors are grateful to Dana Hellman for suggestions on an earlier version of the manuscript.

References

Adger W (2006) Vulnerability. Global Environ Change 16(3):268–281. https://doi.org/10.1016/j.gloenvcha.2006.02.006

Ajibade I, McBean G (2014) Climate extremes and housing rights: a political ecology of impacts, early warning and adaptation constraints in Lagos slum communities. Geoforum 55:76–86. https://doi.org/10.1016/j.geoforum.2014.05.005

Balica SF, Wright NG, van der Meulen F (2012) A flood vulnerability index for coastal cities and its use in assessing climate change impacts. Nat Hazards 64(1):73–105. https://doi.org/10.1007/s11069-012-0234-1

Balstrøm T, Crawford D (2018) Arc-Malstrøm: A 1D hydrologic screening method for stormwater assessments based on geometric networks. Comput Geosci 116:64–73. https://doi.org/10.1016/j.cageo.2018.04.010

Bell ML, O'Neill MS, Ranjit N et al (2008) Vulnerability to heat-related mortality in Latin America: a case-crossover study in São Paulo, Brazil, Santiago, Chile and Mexico City Mexico. Int J Epidemiol 37(4):796–804. https://doi.org/10.1093/ije/dyn094

Berkes F, Colding J, Folke C (2003) Navigating social–ecological systems: building resilience for complexity and change. Cambridge University Press, Cambridge, UK

Berko J, Ingram DD, Saha S et al (2014) Deaths attributed to heat, cold, and other weather events in the United States, 2006–2010. No 76. US Department of Health and Human Services, Centers for Disease Control and Prevention, National Center for Health Statistics

Box GE (1976) Science and statistics. J Am Stat Assoc 71(356):791–799

Bradford K, Abrahams L, Hegglin M et al (2015) A heat vulnerability index and adaptation solutions for Pittsburgh Pennsylvania. Environ Sci Technol 49(19):11303–11311. https://doi.org/10.1021/acs.est.5b03127

Brown K (2012) Policy discourses of resilience. In: Pelling M, Manuel-Navarette D, Redcliff M (eds) Climate change and the crisis of capitalism. Routledge, Abingdon and New York, pp 37–50

Bryant J (2008) Towards delivery and dignity: community struggle from Kennedy Road. J Asian Afr Stud 43(1):41–61. https://doi.org/10.1177/0021909607085585

Carter MR, Little PD, Mogues T et al (2007) Poverty traps and natural disasters in Ethiopia and Honduras. World Dev 35(5):835–856. https://doi.org/10.1016/j.worlddev.2006.09.010

Chow WTL, Chuang W, Gober P (2012) Vulnerability to extreme heat in Metropolitan Phoenix: Spatial, temporal, and demographic dimensions. Prof Geogr 64(2):286–302. https://doi.org/10.1080/00330124.2011.600225
City of New York (2017) Cool neighborhoods NYC. New York City. https://www1.nyc.gov/assets/orr/pdf/Cool_Neighborhoods_NYC_Report.pdf. Accessed 09 Jun 2020
Cutter SL (1996) Vulnerability to environmental hazards. Prog Hum Geogr 20(4):529–539. https://doi.org/10.1177/030913259602000407
Cutter SL, Boruff BJ, Shirley WL (2003) Social vulnerability to environmental hazards. Social Sci Q 84(2):242–261. https://doi.org/10.1111/1540-6237.8402002
Cutter SL, Mitchell JT, Scott MS (2000) Revealing the vulnerability of people and places: a case study of Georgetown County South Carolina. Ann Assoc Am Geogr 90(4):713–737. https://doi.org/10.1111/0004-5608.00219
Hamstead ZA, Farmer C, McPhearson T (2018) Landscape-based extreme heat vulnerability assessment. J Extreme Events 5(4):1850018. https://doi.org/10.1142/S2345737618500185
Hamstead ZA, Kremer P, Larondelle N et al (2016) Classification of the heterogeneous structure of urban landscapes (STURLA) as an indicator of landscape function applied to surface temperature in New York City. Ecol Indic 70:574–585. https://doi.org/10.1016/j.ecolind.2015.10.014
Hattis D, Ogneva-Himmelberger Y, Ratick S (2012) The spatial variability of heat-related mortality in Massachusetts. Appl Geogr 33:45–52. https://doi.org/10.1016/j.apgeog.2011.07.008
Hondula DM, Davis RE, Saha MV et al (2015) Geographic dimensions of heat-related mortality in seven US cities. Environ Res 138:439–452. https://doi.org/10.1016/j.envres.2015.02.033
Inostroza L, Palme M, De Barrera F (2016) A heat vulnerability index: Spatial patterns of exposure, sensitivity and adaptive capacity for Santiago de Chile. PLoS ONE 11(9):1–26. https://doi.org/10.1371/journal.pone.0162464
Intergovernmental Panel on Climate Change (IPCC) (2012) Glossary of terms. In: Field CB, Barros V, Stocker TF et al (eds) Managing the risks of extreme events and disasters to advance climate change adaptation. Cambridge University Press, Cambridge, UK, and New York, Cambridge University Press, pp 555–564
Johnson DP, Stanforth A, Lulla V et al (2012) Developing an applied extreme heat vulnerability index utilizing socioeconomic and environmental data. Appl Geogr 35(1–2):23–31. https://doi.org/10.1016/j.apgeog.2012.04.006
Klinenberg E (2002) Race, place and vulnerability: Urban neighborhoods and the ecology of support. A social autopsy of disaster in Chicago. University of Chicago Press, Chicago, IL, pp 79–128
Kovats RS, Hajat S (2008) Heat stress and public health: a critical review. Annu Rev Public Health 29:41–55. https://doi.org/10.1146/annurev.publhealth.29.020907.090843
Madrigano J, Ito K, Johnson S et al (2015) A case-only study of vulnerability to heat wave–related mortality in New York City (2000–2011). Environ Health Perspect 123(7):672–678. https://doi.org/10.1289/ehp.1408178
Mahanta R, Das D (2017) Flood induced vulnerability to poverty: evidence from Brahmaputra Valley, Assam, India. Int J Disaster Risk Reduct 24:451–461. https://doi.org/10.1016/j.ijdrr.2017.04.014
Medina-Ramón M, Zanobetti A, Cavanagh DP et al (2006) Extreme temperatures and mortality: assessing effect modification by personal characteristics and specific cause of death in a multi-city case-only analysis. Environ Health Perspect 114(9):1331–1336. https://doi.org/10.1289/ehp.9074
Mennis J (2002) Using geographic information systems to create and analyze statistical surfaces of population and risk for environmental justice analysis. Social Sci Q 83(1):281–297. https://doi.org/10.1111/1540-6237.00083
Mennis J (2003) Generating surface models of population using dasymetric mapping. Prof Geogr 55(1):31–42. https://doi.org/10.1111/0033-0124.10042
Morrow BH (1999) Identifying and mapping community vulnerability. Disasters 23(1):1–18. https://doi.org/10.1111/1467-7717.00102

Norris FH, Stevens SP, Pfefferbaum B, Wyche KF, Pfefferbaum RL (2008) Community resilience as a metaphor, theory, set of capacities, and strategy for disaster readiness. Am J Community Psychol 41(1–2):127–150. https://doi.org/10.1007/s10464-007-9156-6
Reid CE, O'Neill MS, Gronlund CJ et al (2009) Mapping community determinants of heat vulnerability. Environ Health Perspect 117(11):1730–1736. https://doi.org/10.1289/ehp.0900683
Rinner C, Rinner C, Patychuk D et al (2010) Vulnerability assessment for the city of Toronto. Cartography Geog Inf Sci 37(1):31–44. https://doi.org/10.1559/152304010790588089
Rosenthal JK, Kinney PL, Metzger KB (2014) Intra-urban vulnerability to heat-related mortality in New York City, 1997–2006. Health Place 30:45–60. https://doi.org/10.1016/j.healthplace.2014.07.014
Sleeter R, Gould M (2007) Geographic information systems software to remodel population data using dasymetric mapping methods. U.S. Geological Survey Techniques and Methods 11-C2
Smargiassi A, Goldberg MS, Plante C et al (2009) Variation of daily warm season mortality as a function of micro-urban heat islands. J Epidemiol Community Health 63(8):659–664. https://doi.org/10.1136/jech.2008.078147
Uejio CK, Wilhelmi OV, Golden JS et al (2011) Intra-urban societal vulnerability to extreme heat: the role of heat exposure and the built environment, socioeconomics, and neighborhood stability. Health Place 17(2):498–507. https://doi.org/10.1016/j.healthplace.2010.12.005
Walker G (2009) Beyond distribution and proximity: exploring the multiple spatialities of environmental justice. Antipode 41(4):614–636. https://doi.org/10.1111/j.1467-8330.2009.00691.x
Wilhelmi OV, Hayden MH (2010) Connecting people and place: a new framework for reducing urban vulnerability to extreme heat. Environ Res Lett 5(1):014021. https://doi.org/10.1088/1748-9326/5/1/014021
Winsemius HC, Jongman B, Veldkamp TIE et al (2018) Disaster risk, climate change, and poverty: assessing the global exposure of poor people to floods and droughts. Environ Dev Econ 23(3):328–348. https://doi.org/10.1017/S1355770X17000444
Wisner B, Blaikie P, Cannon T et al (2003) At risk: natural hazards, people's vulnerability and disasters, 2nd edn. Routledge, Abingdon, United Kingdom

Zoé A. Hamstead is an Assistant Professor of Environmental Planning at the University at Buffalo and Director of the Community Resilience Lab. Her research builds on interdisciplinary approaches in urban planning, geography, and urban ecology to support climate equity and adaptive climate planning approaches. Dr. Hamstead teaches courses in environmental planning, environmental justice, economic concepts, geographic analysis of environmental problems, and engaged community energy studios. She holds a Ph.D. in Urban & Public Policy from The New School, a Master's in City & Regional Planning from the University of North Carolina at Chapel Hill, and a Bachelor of Liberal Arts from St. John's College.

Jason Saueris a Ph.D. candidate in Environmental Life Sciences at Arizo-na State University. His work broadly focuses on methods to determine flood exposure and vulnerability in cities across the US, Mexico, and Chile. His primary site of research is Valdivia, Chile, where he has conducted field and computational research to demonstrate the efficacy of incorporating naturally-occurring wetlands in urban drainage systems in reducing flood exposure during extreme rain events, under present and future climate conditions.

Chapter 5
Producing and Communicating Flood Risk: A Knowledge System Analysis of FEMA Flood Maps in New York City

Robert Hobbins, Tischa A. Muñoz-Erickson, and Clark Miller

Abstract The burgeoning development of coastal cities coupled with increasing exposure to sea level rise and extreme weather events has exacerbated the vulnerability of coastal communities and infrastructure to floods. In order to make good flood risk reduction and resilience decisions, cities are interested in gaining better insights into what are perceived to be the "real" risks of floods. However, what counts as a good estimate of such risks is constructed through the design of a knowledge system that ratifies certain ideas and methods over others. We refer to knowledge systems as the organizational practices and routines that produce, validate and review, communicate, and use knowledge relevant to policy and decision-making. In this chapter, we conduct a knowledge system analysis of FEMA's Flood Insurance Rate Maps in New York City. In 2012, Superstorm Sandy exposed in the national spotlight the shortcomings of how we calculate, map, and use knowledge about flood risk. Through this case study, we hope to demonstrate the value of knowledge systems analysis as a method to stress-test and identify the weaknesses of a knowledge system that warrant attention, as well as to inform potential methods ofupgrading or redesigning that system in support of building resilient cities.

Keywords Knowledge systems analysis · National flood insurance program · Risk communication · Climate resilience

R. Hobbins (✉)
School of Sustainability, Arizona State University, Tempe, AZ, USA
e-mail: rhobbins@asu.edu

T. A. Muñoz-Erickson
USDA Forest Service, International Institute of Tropical Forestry, Río Piedras, PR, USA

C. Miller
School for the Future of Innovation in Society, Arizona State University, Tempe, AZ, USA

Z. A. Hamstead et al. (eds.), *Resilient Urban Futures*, The Urban Book Series,
https://doi.org/10.1007/978-3-030-63131-4_5

5.1 Introduction

The burgeoning development of coastal cities coupled with increasing exposure to sea level rise and extreme weather events has exacerbated the vulnerability of coastal communities and infrastructure to floods. One trillion dollars in United States' coastal assets are currently vulnerable to coastal floods, and sea level rise threatens to expose 13 million people to flooding by 2100 (Reidmiller et al. 2018). Extreme events like Superstorm Sandy have revealed the inadequacies of how we calculate, map, and use knowledge about flood risks. National studies have shown that 25% of Federal Emergency Management Agency (FEMA) flood claims lay outside of the FEMA 100-year flood zone (Blessing et al. 2017). Several studies report that population growth, gross domestic product (GDP), and climate change have all led to significant changes in flood exposure, and estimate that 41 million people—rather than the 13 million people shown in FEMA flood maps—live within the 100-year floodplain (Wing et al. 2018). It is clear that an upgrade, or even a rethinking, is urgently needed in how the United States maps and communicates risks of coastal floods.

In this chapter, we use the knowledge systems analysis framework as a lens to understand the social and technological challenges associated with coastal flood risk analysis, doing so with the objective of informing strategies and innovations needed to overcome those inadequacies. We refer to knowledge systems as the organizational practices and routines that produce, validate and review, communicate, and use knowledge relevant to policy and decision-making (Miller and Muñoz-Erickson 2018; Muñoz-Erickson et al. 2017). Specifically, we conduct a knowledge system analysis of FEMA Flood Insurance Rate Maps (FIRMs) in New York City (NYC)—the largest coastal city of the Urban Resilience to Extremes Sustainability Research Network—to shed light on the social innovations required to make flood risk mapping work better for homeowners, businesses, and cities given our rapidly changing climate and urban landscapes. Cities are interested in improving their understanding of what are perceived to be the "true" or "real" risks of floods, so as to make and inform good decisions. What counts as a good estimate of such risks, however, is constructed through the design of a knowledge system that ratifies certain ideas and methods over others. Through this case study, we demonstrate the value of knowledge systems analysis as a method to stress-test and identify weaknesses and blind-spots that warrant attention. This analysis informs potential solutions to upgrade or redesign that system in support of building resilient cities.

5.1.1 The National Flood Insurance Program

The principal flood risk knowledge system in the United States is the FIRM produced by FEMA's National Flood Insurance Program (NFIP). FIRMs are also known simply as FEMA flood maps. The NFIP is responsible for generating knowledge about flood risk within defined zones, which in turn affects decisions about where and how homeowners and businesses build and the flood insurance rates they pay. The NFIP was created by the National Flood Insurance Act of 1968 and made federal flood insurance available for the first time (Michel-Kerjan 2010). The Flood Disaster Protection Act of 1973 made the purchase of flood insurance mandatory for those living within the boundaries of high-risk zones—the 100-year flood zone as defined by the NFIP (Michel-Kerjan 2010). The initial intent of the program was to provide immediate disaster relief to homeowners after experiencing a flood so they could get back on their feet and move out of the flood zone, ultimately reducing flood risk. Paradoxically, the NFIP instead disincentivized homeowners from moving out of flood-prone areas by shifting the costs to rebuild from the individual to society through heavily subsidized federal flood insurance (Platt 1999). Burby (2006) calls this phenomenon the safe development paradox. Unreliable flood maps (as discussed in this chapter) make this issue even worse when homes in high-risk flood zones are not properly identified and are therefore not required to carry federal flood insurance. As a result, the NFIP does not collect enough insurance premiums to cover its flood claims and has had to rely on tens of billions in government bailouts to remain afloat. Simply put, the NFIP system is broke and broken (Walsh 2017).

There have been several notable reforms to attempt to fix the NFIP. The 1994 Reform Act required FEMA to update its FIRMs every five years, though this policy has not been implemented diligently due to stressed budgets, limited administrative staffing, and appeals processes. The 2009 Department of Homeland Security Appropriations Act required FEMA to modernize flood maps by digitizing hand-drawn maps and updating FIRMs to reflect more recent historical climate data. The digitized maps were to be made publicly available through the FEMA Flood Map Service Center. The 2012 Biggert-Waters Flood Insurance Reform Act (BWFIRA) authorized FEMA to update the FIRM to include the best available scientific data regarding future intensities and frequencies of hurricanes, sea level change, precipitation, and storm surge (Grannis 2012). The BWFIRA attempted to raise insurance rates to reflect a property's "true" risk of flooding once a new flood map or update is produced—effectively eliminating the grandfathering process that was federally subsidizing risky properties with taxpayer money. The grandfathering process prevents owners of homes built before a map update from having to pay the full rate required by a new update. Instead, premiums increase over five years by just 20% per year. There was considerable backlash by flood insurance holders to the BWFIRA primarily due to the discontinuation of grandfathering. This political battle resulted in two additional bills which rolled back key provisions in the BWFIRA. The Consolidated Appropriations Act of 2014 prohibited FEMA from implementing Section 207 of the BWFIRA, which directed FEMA to use insurance rates commensurate with their full risk after a FIRM update. The 2014 Homeowner Flood Insurance Affordability Act restored the practice of grandfathering.

5.1.2 *Flood Insurance Rate Maps as a Knowledge System*

Flood zones are demarcated by FEMA through a highly routinized process. Professional engineers use hydrodynamic modeling to calculate the expected height (i.e., base flood elevation or BFE) and location of floods by waterbodies such as rivers and oceans; the models do not consider floods from infrastructure failures, pluvial floods, or groundwater sources. For inland areas, flood zones and BFE are determined by modeling the overflow of water from streams that have exceeded their capacity during intense precipitation events (i.e., fluvial floods). In coastal areas, flood zones and the BFE are determined by several parameters: current sea level, wave setup, normal high tide, storm surge, and wave effects. Both fluvial and coastal flood modeling utilize digital elevation models (DEM)—typically derived from light detection and ranging (LiDAR) data—for determining the elevation profiles of the study area. The special flood hazard area (SFHA)—for both inland and coastal areas—is defined as the area exposed to a 1% annual exceedance probability (AEP) of experiencing a flood in any given year. This area is often referred to interchangeably by its return period—the amount of time between floods of a certain size. A flood with T year return period will have a 1/T probability of occurring in any given year (Lin et al. 2012; McPhillips et al. 2018). As such, the return period for an AEP of 1% would be 100 years and the storm would be called a 100-year storm. The 100-year storm standard was selected as a compromise between two competing values: minimizing loss of life by restricting development in floodplains, and keeping floodplains open for economic and urban development (FEMA 2019a). The AEP is determined using statistical frequency analysis of past storms using historical weather data for fluvial floods, and synthetic storms (created from historical storm surge and tidal records, coastline profiles, and simulated laws of physics) for coastal floods (Sobel 2014). The SFHA determines the areas where flood insurance is required and where to enforce floodplain regulations. In addition to the SFHA, flood maps include the areas exposed to a 0.2% AEP storm (i.e., 500-year flood) for reference only. The teal- and black-dotted zones on a FIRM demarcate the 100-year and 500-year flood zones, respectively (see Fig. 5.1). A common criticism of this system is that flood risk for a property is often misconstrued as binary—a property is either in a flood zone or out of it (Kousky 2018). The 500-year flood zone line on flood maps creates this false sense of security on the other side of that line. To make matters worse, FEMA's terminology of a 100-year or 500-year flood zone is also misinterpreted by those who are actually aware that they are in one of those flood zones. For those living in a 100-year flood zone, the message received is that their property will only flood once in 100 years when, in reality, FEMA is trying to communicate that the risk is a 1% probability of flooding every year (FEMA 2017). For instance, over the course of a 30-year mortgage, a property has a 26% chance of flooding. However, as shown throughout this chapter, that is not the "real" risk either.

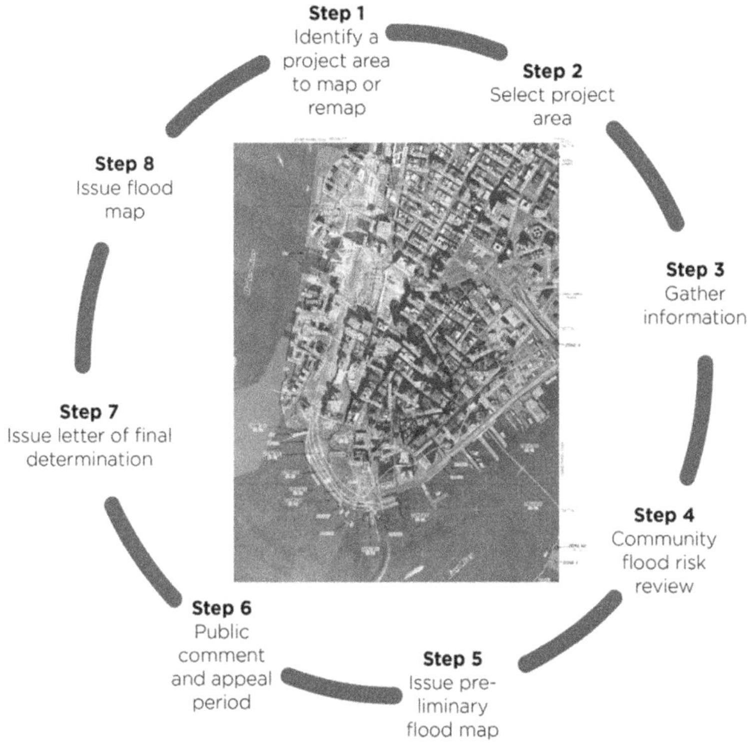

Fig. 5.1 The process for creating a Federal Emergency Management Agency (FEMA) flood map. The current regulatory FEMA Flood Insurance Rate Map for lower Manhattan is shown in the center of the figure. Adapted from FEMA (2019a). Lower Manhattan FIRM courtesy of the FEMA Flood Map Service Center (FEMA nd)

FEMA flood maps are the product of an eight-step iterative process (Fig. 5.1) that begins by identifying a project area (Step 1), deciding on a watershed to map or remap (Step 2), and gathering technical data such as hydrological, infrastructural, land use, and population data (Step 3). A Flood Insurance Study is produced and then shared with community officials to review and provide feedback (Step 4). Once the preliminary FIRM is issued (Step 5), the FIRM can be amended or revised through individual or community appeals (Step 6; FEMA 2019b). Individual property owners can submit a Letter of Map Amendment to provide data showing that their property is not within the SFHA. Community officials can submit a Letter of Map Revision (LOMR) using new scientific or technical data to revise a flood map. Both the LOMA and the LOMR do not actually lead to a physically revised flood map—the changes are documented in letter format only. The Chief Executive Officer (CEO) of a community is the only person who can submit a Physical Map Revision (PMR) to FEMA to physically change the flood zones on a FIRM. Both the PMR and LOMR are typically prepared by experts contracted by local governments. As such, these revisions are costly and resource-intensive endeavors. Once the appeals period expires, a letter of final determination is sent to notify the CEO that the community has six months to adopt a compliant floodplain management ordinance (Step 7) before the new regulatory FIRM becomes effective (Step 8). The case study presented in this chapter will analyze the production, revision, validation, communication, and use of FEMA maps in NYC since 1981.

5.1.3 Knowledge Systems Analysis

looseness-1Knowledge systems analysis is a useful framework to explore the underlying ideas, rationales, social practices, and institutional structures that define sustainability, resilience, and environmental problems. The framework has been applied to analyze a variety of socio-environmental issues, including sustainability visions (Muñoz-Erickson 2014), green infrastructure (Matsler 2017), cloudburst flood resilience (Rosenzweig et al. 2019), integration of citizen and technical flood risk knowledge (Ramsey et al. 2019), and the scalar politics of coastal flood risk (Rozance et al. 2019).

Like systems in general, knowledge systems are described in terms of the functions, elements, and complexities of the systems (Miller and Muñoz-Erickson 2018). The core functions of a knowledge system include the production, validation, review, communication, and use of knowledge. For our FEMA case, the process of developing the FEMA flood map is what defines this knowledge system. The steps shown in Fig. 5.1 reflect the various actors involved in how this knowledge system works, including the production of the flood map by FEMA engineers and city leaders (Steps 1 to 3), the review and validation of the maps by local community leaders (Steps 4, 6, and 7), its communication through the issuing of the preliminary FIRM (Step 5) and regulatory FIRM (Step 8), and its use in decision-making processes as to where to build, how high to build, and what flood insurance rates to charge homeowners. Elements of a knowledge system include the content of that knowledge (including its associated uncertainties), the values embedded in that knowledge, the epistemologies (or how we know what we know), and the institutional structures (people and organizations) through which knowledge is constructed and put to use. For the FIRM, knowledge consists of the actual flood maps that are produced and the knowledge claims that are made regarding those maps (e.g., homes in the FEMA 100-year flood zone have a 1% rate of flooding in any given year). Values may include how the knowledge system prioritizes urban and economic development versus restricting development in flood zones, decisions to set risk boundaries in terms of specific flood return periods (e.g., 100-year and 500-year flood zones), and decisions about how to balance historical data and future projections in setting risk zone boundaries. Epistemologies refer to how the problem is framed, types of evidence (e.g., rainfall data from the past 50 years, LiDAR satellite data, etc.), and the information technologies (e.g., hydrological models) used to produce flood maps. Structures include actors or stakeholders that are involved in the functions of the knowledge system. Analyzing knowledge system structures often reveals how power and authority are distributed and the consequences that these arrangements have on the production, communication, and use of knowledge (Muñoz-Erickson and Cutts 2016; Muñoz-Erickson et al. 2017; Ramsey et al. 2019). The role of power and authority in the operations of the FEMA flood map knowledge system in NYC will also be explored in the next section.

5.2 New York City Flood Map Case Study

> Our city needs precise flood maps that reflect real risks, both today and years from now, and we have to do that fairly—NYC Mayor Bill de Blasio

To conduct the knowledge system analysis of FEMA flood maps for the NYC case study, we use the framework outlined above to review official FEMA products and documents, reports, academic publications, and newspaper articles containing accounts by various types of flood map users. The above quote by Mayor de Blasio highlights the main aspirations and challenges with flood risk mapping in NYC and the nation. City governments value accurate maps that reflect the "real" risks of floods and communicate reliable information about future flood risk to the public. Yet, city governments also wish to have this risk analysis done in a way that does not place unnecessary burdens on homeowners (e.g., higher insurance premiums or decreased home values) or slow down local economic growth (due to restrictions on development in ever-expanding flood zones). The technical flood mapping process is performed within this negotiation of values and risk tolerance. As such, flood maps are more than just technical products—they are maps with great social implications that warrant care in how they are produced so as to not disproportionally or inappropriately impact any particular social group or sector. At the same time, many hurdles must be overcome in efforts to include future flood risks into FEMA flood map products due to the large uncertainties inherent in future climate and sea-level projections. Through this case study, we use knowledge system analysis to illustrate both the technical and socio-political processes—spanning almost four decades (see Fig. 5.2)—that went into the production, validation, communication, and use of FIRMs in NYC, and the implications this has for resilience to extreme flood events.

Superstorm Sandy, which made landfall in NYC on Oct 29, 2012, was one of the worst natural disasters the city has experienced. Sandy was responsible for $19 billion in losses and 43 deaths throughout New York, as well as $65 billion in losses and 159 deaths nationwide (PlaNYC 2013). Sandy's storm surge of 14 + feet (ft) left parts of NYC in ruins and nearly two million residents without power for up to two weeks (PlaNYC 2013).

The damage from Sandy resulted from a storm surge that was the highest in the historical tide gauge record—extending as far back as 1850—and exacerbated by a seasonal high tide that inundated areas well beyond FEMA flood zones. As seen in Fig. 5.3, sea level rise also played a small but significant role in contributing to

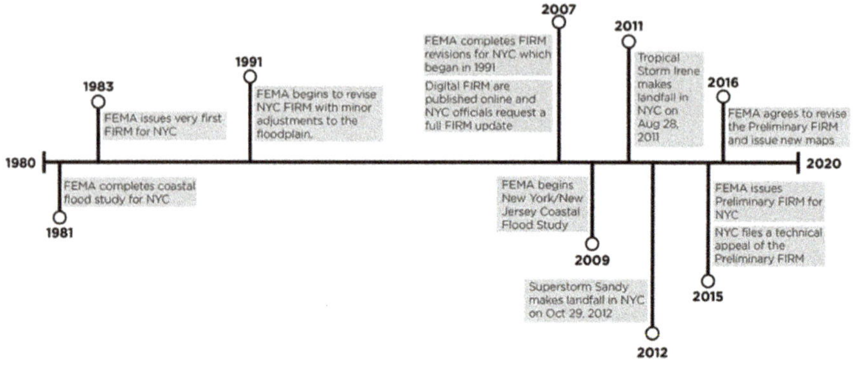

Fig. 5.2 Timeline of Federal Emergency Management Agency flood map production for New York City. Adapted from PlaNYC (2013)

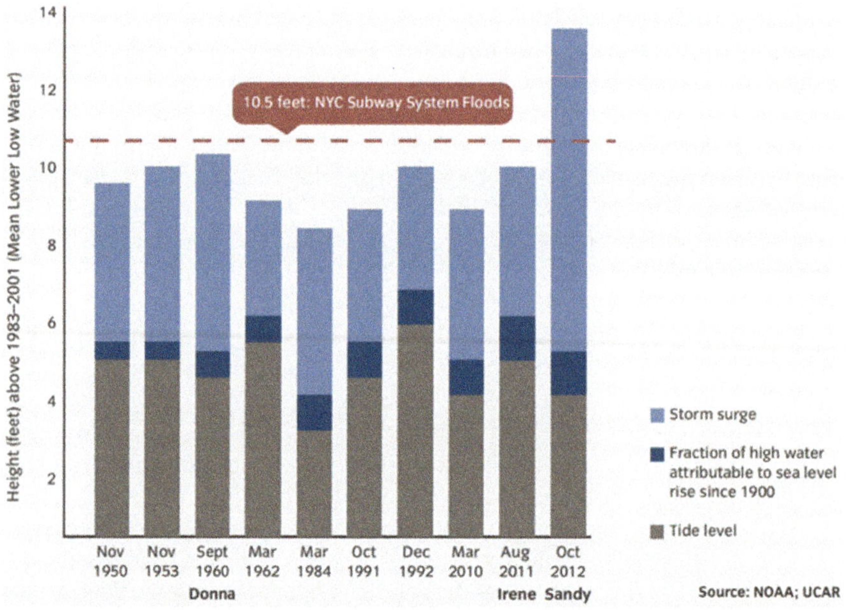

Fig. 5.3 Historical high-water events in lower Manhattan. Used with permission of the New York City Department of City Planning.

the record flood height. At the time of Sandy's landfall, the flood maps were grossly outdated—they did not reflect changes in climate and sea levels (see Fig. 5.4), rapid land-use change, or advances in technology such as the development of more accurate elevation profiles from LiDAR (Parris 2014).

The regulatory flood maps for NYC have not received a significant update since 1983, despite the legal requirement for flood maps to be updated every five years.

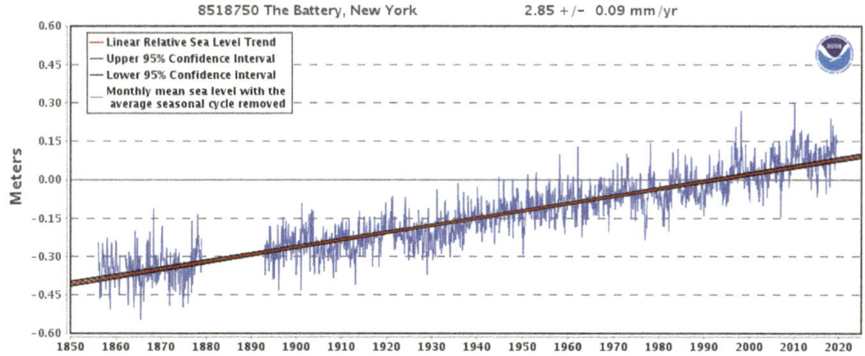

Fig. 5.4 Relative sea level trend as measured from The Battery tide gauge station in NYC. Plotted values are monthly averages. The historic rate of sea level rise is 2.85 mm/year, or about 1 foot every 100 years (National Oceanic and Atmospheric Administration 2019)

From 1991 to 2007, flood map updates included new wetland and stream modeling but failed to include any elevation adjustments. In effect, these were very minor modifications to the original 1983 floodplains. The results were placed on satellite imagery, digitized, and made available online for general public consumption in 2007. Concerned about the inaccurate flood risk information being communicated to the public, local and state officials immediately called on FEMA to perform a full map update using the best scientific data and technology available. The update process did not begin until 2009 and had yet to be completed before Sandy struck in October of 2012 (see Fig. 5.2 timeline).

The 2007 FIRM underestimated the scope of inundation that awaited the city during Sandy. Only 54 and 47% of the flooded area in Queens and Kings, respectively, was predicted by the 1983 flood maps during Sandy (Shaw et al. 2013). Figure 5.5 shows the vast swaths of the city inundated by Sandy, yet left out from the 1983 FIRM 100-year floodplain. However, Sandy was not calculated to be a 100-year storm; it was estimated by using outdated historical climate data to be a 1,000-year storm (Lin et al. 2012). However, several authors argue that climate change helped to intensify Superstorm Sandy (Dietrich 2017; Parris 2014; Sobel 2014). Increases in sea levels alone could have accounted for half a foot of flooding during Sandy (Parris 2014; Shaw et al. 2013). Lin et al. (2012) show that when taking into consideration changing climate and increasing sea levels, the current 100-year storm surge event in NYC has the potential to occur every 20 years or less and the present 500-year event has the potential to occur every 240 years or less by 2100. Thus, there are strong reasons to update flood maps regularly to reflect changing climate and sea levels. If the FEMA flood maps had been updated prior to Sandy to incorporate recent SLR and extreme precipitation and flooding events (e.g., the March 2010 nor'easter and Tropical Storm Irene in 2011), they may have more accurately reflected the extent of flood risk during Sandy and improved flood risk communication and resilience outcomes.

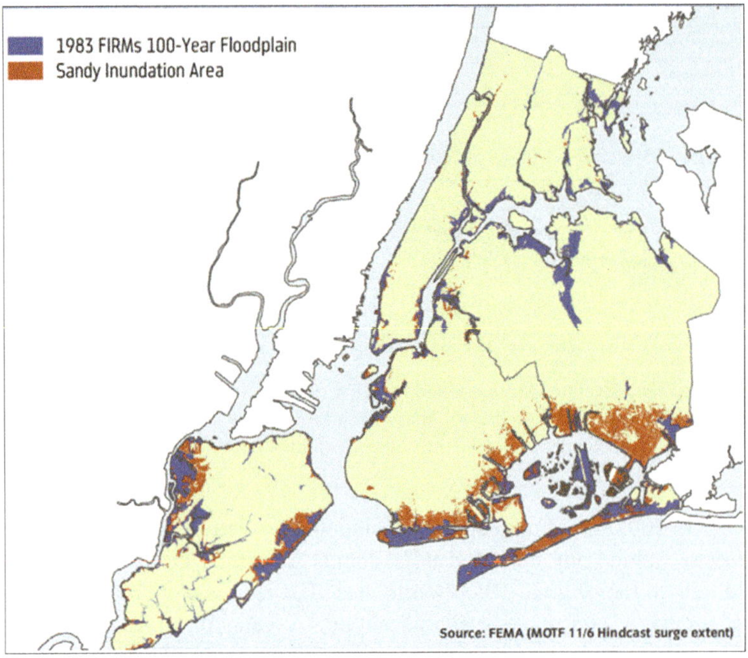

Fig. 5.5 1983 federal emergency management agency flood insurance rate map and Sandy inundation area comparison (PlanNYC 2013). Image used with permission of the New York City Department of City Planning. All rights reserved

After completing the Coastal Flood Study for New York in 2009, FEMA issued the 2015 Preliminary FIRM (P-FIRM) for NYC using new LiDAR data, more recent climatological data (e.g., Tropical Storm Irene and Superstorm Sandy were both included), and more sophisticated hydrologic modeling. The 2015 P-FIRM nearly doubled the building stock located in the 100-year flood zone from 36,000 to 71,500 units (City of New York Mayor's Office of Recovery and Resiliency 2015). Nearly twice as many New Yorkers would be required to pay for mandatory flood insurance after this update. The P-FIRM had the potential to aggravate the affordable housing crisis in NYC by expanding the reach of mandatory flood insurance and increasing existing premiums (Dixon et al. 2017). Consequently, the news was not welcomed by affected homeowners (Chen 2018). Under public pressure to keep housing and insurance rates affordable, NYC pushed back by filing an appeal of the 2015 P-FIRM on scientific and technical grounds (Chen 2018). The City's appeal was politically motivated, but had to be filed on scientific and/or technical grounds—FEMS's epistemology for creating and revising flood maps. As discussed in the section entitled "Flood Insurance Rate Maps as Knowledge Systems," the Chief Executive Officer of a community has the sole legal authority to challenge FEMA's flood mapping expertise. The appeal must also be submitted within a 90-day period after a P-FIRM is issued. The New York City Mayor's Office contracted outside engineering firms, which included the design and consultancy firm Arcadis, to conduct the City's flood analysis. NYC's appeal claimed that scientific and technical errors—insufficient extratropical storm model validation and misrepresentation of tidal effects of extratropical storms—lead to the P-FIRM overstating the BFE by over 2 ft in many areas and presenting 35% larger SFHA boundaries (City of New York Mayor's Office of Recovery and Resiliency 2015). However, NYC elsewhere claimed that the initial reason for the appeal was that "the revisions will assist New York City in making coastlines more resilient and climate ready, while ensuring homeowners are not required to purchase more insurance than their current flood risk requires" (City of New York nd). The appeal was an attempt to reduce the extent of the new SFHA and BFE in the P-FIRM (the political goal) while also updating the maps with more recent climate and storm data (the resiliency goal). Rather than 71,500 buildings in the SFHA, the new NYC analysis reduced the number of units to just 45,000—a 37% reduction—as shown on the P-FIRM. The appeal also provided extra time before an update could be issued—effectively saving property owners money as their insurance rates and requirement to purchase flood insurance would continue to be based on the 2007 FIRM SFHA boundaries. The City's appeal was successful. FEMA agreed in 2016 to revise the maps according to the City's analysis. However, as of December 2019, FEMA has still not issued any update to NYC's FIRM. As such, there are now three competing knowledge claims regarding claims regarding New Yorkers' FEMA-delineated flood risk, leaving residents in limbo regarding this risk (e.g., the current regulatory 2007 FIRM, the 2015 Preliminary FIRM, and NYC's flood analysis). While the City's political goal may have been achieved through this appeal, this state of uncertainty is a failure of the flood mapping knowledge system to clearly, timely, and definitively communicate flood risk to property owners for their individual resilience and adaptation decisions. For instance, a prospective homebuyer may unknowingly become vulnerable to floods by purchasing a new home that is within the SFHA on NYC's flood analysis, but does not fall within this zone according to the 2007 FIRM—the map currently used to determine flood risk for a property. For instance, many residents of Staten Island—one of the hardest hit places during Sandy—expressed frustration that they did not know their properties were at risk of flooding at the time they purchased their homes (Moore 2018). The Morgan family—whose basement was destroyed in Sandy—said they would have at least moved their utilities out of the basement had they known Sandy was predicted to bring 11 ft of flooding—as shown on the P-FIRM—compared to the less than 1 ft shown on the 2007 FIRM (Shaw et al. 2013).

In contrast, there is actually a clear and definitive standard for resolving these competing flood risk knowledge claims for use in building construction at the city level. NYC adopted Local Law 96/13 which modified the City's building code to require all work permits for construction projects to be based on the more restrictive BFE and SFHA of either the 2007 FIRM or the P-FIRM (NYC Buildings 2014). Additionally, the NYC Commissioner of Buildings issued a rule in 2013 that for buildings in the SFHA, 1 to 2 ft must be added to the BFE in order to determine the Design Flood Elevation (DFE). No dwelling units or mechanical equipment (e.g., electrical and HVAC systems) are permitted in floors below the DFE (New York City

Planning Department 2013). By decoupling the P-FIRM from insurance rate hikes, NYC was able to make use of this valuable knowledge for construction decisions without imposing new or higher flood insurance costs on residents.

While the P-FIRM and NYC's flood analysis incorporated more recent climate data, these maps still do not incorporate any anticipated future flood risk (e.g., sea level rise) for long-term residential or urban planning decisions. NYC addressed this knowledge gap in 2008 by creating a new knowledge system separate from the NFIP. The New York Panel on Climate Change (NPCC) is a panel of experts created by the NYC Mayor's Office to provide analysis of future climate change impacts such as extreme floods. FEMA is now collaborating with the NPCC to create "innovative, climate-smart flood maps" for NYC that incorporate the best available science regarding future sea levels and coastal storms for long-term planning and building purposes, while updating the FIRM to depict current risk for insurance purposes (FEMA 2016). The NPCC recently published its projections of NYC's floodplain for 2100 and compared it to the 2015 P-FIRM (Patrick et al. 2019). The results indicate that the floodplain is likely to expand as NYC experiences additional sea level rise and more intense storm surges (Fig. 5.6).

The NPCC's anticipatory flood maps are not yet required for NYC's long-term planning decisions, but the City now has access to this valuable knowledge. While the NPCC has been helpful for the City to understand their future flood risk, individual New Yorkers are still largely in the dark. NYC has recently created a new position, Deputy Director of Climate Science and Risk Communication, to serve as a City liaison to the NPCC. There is hope that the creation of this new position may help communicate the NPCC's forward-looking flood risk maps to the general public.

The strategy of decoupling flood risk knowledge from insurance rates is at the core of this knowledge innovation for anticipatory flood resilience decision-making in NYC. Access to resources—money and experts—were also essential. NYC had the resources to convene the expert NPCC panel to produce this knowledge for the City's planning and decision-making. Yet, few cities have NYC's financial and university resources to be able to create an entirely new knowledge system—such as the NPCC—to augment the inadequate FEMA flood maps. From a social justice and equity perspective, it is important that FEMA step in to provide access to future-looking flood risk knowledge for resource-scarce cities. However, there is not a clear path forward for how FEMA will communicate future risks of flooding for community resilience and adaptation decisions. FEMA has been authorized to provide maps of future flood risk since the BWFIRA was enacted in 2012. However, the FEMA Technical Mapping Advisory Council's efforts have been stalled and their final report withheld, preventing legally binding guidance on how FEMA should move forward with communicating future flood risks. In the following section, we discuss some possible options for redesigning the NFIP based on this knowledge system analysis of NYC flood risk mapping.

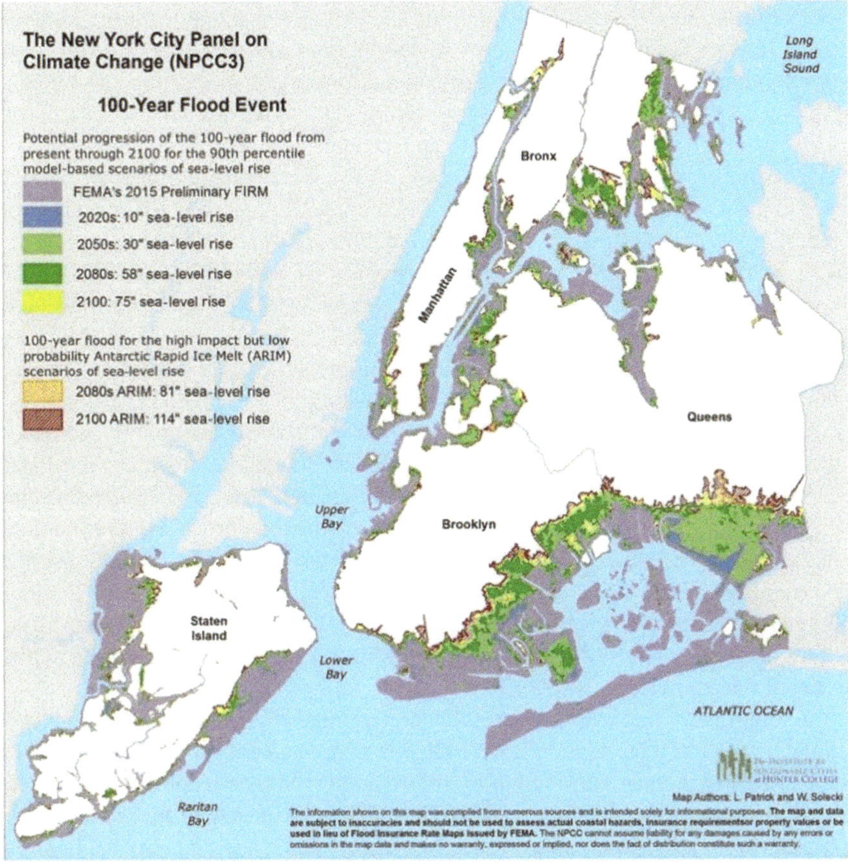

Fig. 5.6 Projected 100-year floodplain through 2100, as compared to the 2015 Preliminary Flood Insurance Rate Map (Patrick et al. 2019)

5.3 Discussion and Conclusion

Understanding how the FEMA flood map knowledge system works is essential for the adaptive capacity and resilience of cities to climate change and extreme events. These maps guide a myriad of important decisions affecting urban form and community resilience both now and in the long-term future. Homeowners use this information to make individual decisions such as whether to buy a home, carry flood insurance for a home, how high to elevate a home, or simply whether or not to move a generator or other appliances out of their basement or ground floor. Developers use this information to decide where to build and the design of the building. City engineers use this information to determine where and how to build critical infrastructure throughout the city. As the U.S. Department of Homeland Security Office of the

Inspector General (DHS OIG 2017) reported, it is imperative that we provide accurate and reliable flood risk information to the public, and this will require changes to the flood mapping process, management, and oversight. In essence, the DHS is calling for a knowledge system upgrade or redesign to modernize the flood mapping process given its expanded user network and salience.

As we have shown with the NYC case study, the FEMA flood map knowledge system has several social-political and technical challenges associated with it, including outdated climate data, lack of anticipatory flood risk knowledge, difficult to interpret and communicate flood risks, lack of consideration of infrastructure or pluvial floods, politically motivated map revisions, a resource-intensive and inequitable revision process, and so on. How well a knowledge system produces quality knowledge for decision-making is not simply a matter of collecting the best scientific data and using the most sophisticated technology to produce a flood map; the distribution of power and authority also greatly influences the quality and accuracy of the knowledge claims produced by the knowledge system (e.g., the SFHA boundaries and BFE of the P-FIRM and NYC flood analyses). In NYC, the social (e.g., the formalized and routinized process of creating map products) and political (e.g., who wins and who loses from map updates, who has authority to challenge flood map knowledge claims, etc.) dynamics have played key roles in the production, review and validation, communication, and use of flood maps over the past four decades. Any redesign will need to address both the social-political and technical aspects of this knowledge system.

You might ask, what would an upgraded or redesigned flood mapping system look like and how could it be accomplished? The low-hanging fruit for an upgrade would be for FEMA to include non-regulatory future flood risk knowledge alongside their official regulatory map products; this would effectively decouple this information from determining insurance rates. As shown in the NYC case study, by decoupling the P-FIRM from insurance rates, NYC was able to use this valuable knowledge for building construction and zoning decisions to improve the long-term flood resilience of the City's built environment. A more transformative change to the entire flood mapping system would be to retire the use of the 100-year and 500-year flood zones given the well-documented misconceptions users have and the false sense of security they give to residents living outside of these zones. This technical change will also be inherently disruptive socio-politically as new federal legislation would need to be written and the entire NFIP—which provides disaster relief to flood victims—would need to be dramatically revised to accommodate this change. This redesign would likely require new legislation from the U.S. Congress. It would also likely require a shift in the values underpinning the knowledge system—which are notoriously difficult to change. Given the magnitude of recent flood disasters like Hurricane Katrina, Superstorm Sandy, Hurricane Harvey, and Hurricane María, it may become necessary to value the protection of lives and property more than is currently done relative to the value accorded to urban development and growth. The Special Hazard Flood Area—which restricts development in the 100-year flood zone—was chosen as a balance between these two values. The NFIP may require a recalibration of our nation's flood risk tolerance and values in order to fix this broken and broke program.

In closing, our analysis of how the FEMA FIRM knowledge system works sheds light on the underlying complex social and political dynamics involved in how we know, review and validate, communicate, and use flood risk knowledge. Knowledge about flood risks is more than the map that results from collecting data and running models to determine "real" flood risk for a property; it is the outcome of a highly contested co-production process between individual residents, experts (e.g., engineers and hydrologists), city officials, federal government agencies, and other stakeholders as they seek to map flood risk while trying to achieve their diverse and conflicting goals (e.g., minimizing flood insurance costs while improving the accuracy of flood maps). Many important technological innovations are being developed to improve how we calculate flood risks, including, for instance, advances in real-time flood sensor systems, sophisticated hydrological models, and use of high-resolution satellite data. These innovations will fall short, however, if they don't also address the non-technical and social aspects crucial to making knowledge systems work. In light of accelerated climate change and extreme coastal events, we suggest that more attention toward understanding flood risk as a knowledge systems problem can further advance resiliency goals for coastal cities.

Acknowledgements This material is based upon work supported by the National Science Foundation-funded Urban Resilience to Extreme Weather-Related Events Sustainability Research Network (UREx SRN; NSF grant no. SES 1444755), as well as dissertation research grants provided to Robert Hobbins from the Graduate Professional Student Association at Arizona State University and the American Association of Geographers. The authors are also very grateful for the time and information provided by practitioners in our case study cities and from the two anonymous reviewers who provided valuable feedback on earlier versions of this chapter.

References

Blessing R, Sebastian A, Brody SD (2017) Flood risk delineation in the U.S.: How much loss are we capturing? Nat Hazard Rev 18(3):1–10. https://doi.org/10.1061/(ASCE)NH.1527-6996.0000242

Burby RJ (2006) Hurricane Katrina and the paradoxes of government disaster policy: Bringing about wise governmental decisions for hazardous areas. Ann Am Acad Political Soc Sci 604(1):171–191. https://doi.org/10.1177/0002716205284676

Chen DW (2018) In New York, drawing flood maps is a "game of inches." https://www.nytimes.com/2018/01/07/nyregion/new-york-city-flood-maps-fema.html. Accessed 01 Nov 2019

City of New York Mayor's Office of Recovery and Resiliency (2015) Appeal of FEMA's preliminary flood insurance rate maps for New York City. https://www1.nyc.gov/assets/floodmaps/images/content/pages/1-NYC%20FEMA%20Appeal%20FINAL%20with%20Appendices%20and%20Cover%20Letter%2006252015_web.pdf. Accessed 15 Dec 2019

City of New York (nd) Appeals. NYCFlood Maps. https://www1.nyc.gov/site/floodmaps/appeals/overview.page. Accessed 11 Nov 2019

Dietrich JC (2017) Vignette: climate change effects on flooding during Hurricane Sandy. In: Horney J (ed) Disaster epidemiology: methods and applications, 1st edn. Academic Press, Cambridge, MA, pp 153–156

Dixon L, Clancey N, Miller B et al (2017) The cost and affordability of flood insurance in New York City: economic impacts of rising premiums and policy options for one- to four-family homes. https://www.rand.org/pubs/research_reports/RR1776.html. Accessed 14 Dec 2019

Federal Emergency Management Agency (2016) Mayor De Blasio and FEMA announce plan to revise NYC'sflood maps. https://www.fema.gov/news-release/2016/10/17/mayor-de-blasio-and-fema-announce-plan-revise-nycs-flood-maps. Accessed 11 August 2019

Federal Emergency Management Agency (2017) FEMA fact sheet: common questions about flood maps and risk. https://www.fema.gov/media-library-data/1513200364180-bab79b0ae4855f62ebc94baa06fc0186/Common_Questions_Fact_Sheet.pdf. Accessed 14 Dec 2019

Federal Emergency Management Agency (2019a) The Risk MAP flood riskproject lifecycle. Department of Homeland Security. https://www.fema.gov/risk-map-flood-risk-project-lifecycle. Accessed 14 Dec 2019

Federal Emergency Management Agency (2019b) Flood map revision process. https://www.fema.gov/flood-map-revision-processes. Accessed 14 Dec 2019

Federal Emergency Management Agency (nd) FEMA flood map service center. Department of Homeland Security. https://msc.fema.gov/portal/search?AddressQuery=lower%20manhattan#searchresultsanchor. Accessed 14 Dec 2019

Grannis J (2012) Analysis of how the flood insurance reform act of 2012 (H.R. 4348) may affect state and local adaptation efforts. Georgetown Climate Center. https://www.georgetownclimate.org/files/report/Analysis%20of%20the%20Flood%20Insurance%20Reform%20Act%20of%202012.pdf. Accessed 11 Aug 2019

Kousky C (2018) How America fails at communicating flood risks. City Lab. https://www.citylab.com/environment/2018/10/how-america-fails-communicating-flood-risks/572620/. Accessed 11 Aug 2019

Lin N, Emanuel K, Oppenheimer M et al (2012) Physically based assessment of hurricane surge threat under climate change. Nat Clim Change 2(6):462–467. https://doi.org/10.1038/nclimate1389

Matsler AM (2017) Knowing nature in the city: comparative analysis of knowledge systems challenges along the eco-techno spectrum of green infrastructurein Portland & Baltimore. Dissertation, Portland State University

McPhillips LE, Chang H, Chester M et al (2018) Defining extreme events: a cross-disciplinary review. Earth's Future 6(3):441–455. https://doi.org/10.1002/2017EF000686

Michel-Kerjan EO (2010) Catastrophe economics: the national flood insurance program. J Econ Perspect 24(4):165–186. https://doi.org/10.1257/jep.24.4.165

Miller C, Muñoz-Erickson T (2018) The rightful place of science: designing knowledge. Consortium for Science, Policy & Outcomes, Phoenix

Moore R (2018) For Sandy survivors this program made all the difference. https://www.nrdc.org/experts/rob-moore/title. Accessed 14 Dec 2019

Muñoz-Erickson TA (2014) Multiple pathways to sustainability in the city: The case of San Juan, Puerto Rico. Ecol Soc 19(3). https://doi.org/10.5751/ES-06457-190302

Muñoz-Erickson TA, Cutts BB (2016) Structural dimensions of knowledge-action networks for sustainability. Curr Opin Environ Sustain 18:56–64. https://doi.org/10.1016/j.cosust.2015.08.013

Muñoz-Erickson TA, Miller CA, Miller TR (2017) How cities think: knowledge co-production for urban sustainability and resilience. Forests 8(6):1–17. https://doi.org/10.3390/f8060203

National Oceanic and Atmospheric Administration (2019) Relative Sea Level Trend, 8518750 The Battery, New York. NOAA Tides and Currents. https://tidesandcurrents.noaa.gov/sltrends/sltrends_station.shtml?id=8518750. Accessed 13 Nov 2019

New York CityPlanning Department (2013) Coastal climate resilience: Designing for flood risk. www.nyc.gov/designingforfloodrisk. Accessed 11 Nov 2019

NYCBuildings (2014) Recently enacted resiliency legislation. https://www1.nyc.gov/assets/buildings/pdf/summary_resiliency_legislation.pdf. Accessed 11 Nov 2019

Parris A (2014) How Hurricane Sandy tamed the bureaucracy. Issues Sci Technol 30(4):83–90

Patrick L, Solecki W, Gornitz Vet al (2019) Chapter 5: mapping climate risk. In: Rosenzweig C, Solecki W (eds) New York City panel on climate change2019 report. Annals of the New York Academy of Science, New York City, pp 115–125. https://doi.org/10.1111/nyas.14015
PlaNYC (2013) A stronger, more resilient NYC. City of New York. https://www1.nyc.gov/site/sirr/report/report.page. Accessed 14 Aug 2019
Platt RH (1999) Disasters and democracy: the politics of extreme natural events, 2nd edn. Island Press, Washington, DC
Ramsey MM, Muñoz-Erickson TA, Mélendez-Ackerman E et al (2019) Overcoming barriers to knowledge integration for urban resilience: a knowledge systems analysis of two flood prone communities in San Juan, Puerto Rico. Environ Sci Policy 99:48–57. https://doi.org/10.1016/j.envsci.2019.04.013
Reidmiller DR, Avery CW, Easterling, DR et al (2018) Fourth national climate assessment. https://nca2018.globalchange.gov/chapter/front-matter-about/. Accessed 11 Aug 2019
Rosenzweig B, Ruddell BL, McPhillips L et al (2019) Developing knowledge systems for urban resilience to cloudburst rain events. Environ Sci Policy 99:150–159. https://doi.org/10.1016/j.envsci.2019.05.020
Rozance MA, Denton A, Matsler M (2019) Examining the scalar knowledge politics of risk within coastal sea level rise adaptation planning knowledge systems. Environ Sci Policy 99:105–114. https://doi.org/10.1016/j.envsci.2019.05.024
Shaw A, Thompson C, Meyer T (2013) Federal flood maps left New York unprepared for Sandy—and FEMA knew it. ProPublica. https://www.propublica.org/article/federal-flood-maps-left-new-york-unprepared-for-sandy-and-fema-knew-it. Accessed 11 Nov 2019
Sobel AH (2014) Storm surge. HarperCollins, New York
U.S. Department of Homeland Security Office of the Inspector General (2017) FEMA needs to improve management of its flood mapping programs. https://www.oig.dhs.gov/sites/default/files/assets/2017/OIG-17-110-Sep17.pdf. Accessed 11 Aug 2019
Walsh MW (2017) A Broke, and broken, flood insurance program. New York Times. https://www.nytimes.com/2017/11/04/business/a-broke-and-broken-flood-insurance-program.html. Accessed 11 Aug 2019
Wing OEJ, Bates PD, Smith AM et al (2018) Estimates of present and future flood risk in the conterminous United States. Environ Res Lett 13(034023):1–7. https://doi.org/10.1088/1748-9326/aaac65

Robert Hobbins is a Ph.D. Candidate at the School of Sustainability at Arizona State University and an NSF Research Intern at the USDA Forest Service International Institute of Tropical Forestry. Robert has a BS in Physics and a Master of Education from The Ohio State University. He also earned an MS in Community Resources & Development—Sustainable Communities at Arizona State University. Robert joined the Urban Resilience to Extremes Sustainability Research Network (UREx SRN) as a Graduate Fellow in 2016 and continues to be active in the network by supporting the planning and implementation of scenario workshops and working with UREx SRN city practitioners to build local capacities to transition to more sustainable and resilient urban futures.

Tischa A. Muñoz-Erickson is a Research Social Scientist in the USDA Forest Service's International Institute of Tropical Forestry, in Río Piedras, Puerto Rico. She studies urban sustainability governance, including the policy networks, knowledge systems, anticipatory capacities, and strategies to advance sustainability, resilience, and equity. She is also actively involved in transdisciplinary platforms to facilitate the co-production of futures and transition pathways in the US and Latin American cities. Muñoz-Erickson leads the International Urban Field Station and the San Juan ULTRA and is also Co-PI of the National Science Foundation Urban Resilience to Extreme Events Sustainability Research Network (UREx SRN) and the SETS Convergence project.

Clark Miller is Professor and Director of the Center for Energy and Society in the School for the Future of Innovation in Society at Arizona State University. His work offers theoretical, practical, and critical insights into how societies can better design and deploy knowledge and technology in the service of more just and sustainable futures.

Chapter 6
Positive Futures

David M. Iwaniec, Marta Berbés-Blázquez, Elizabeth M. Cook, Nancy B. Grimm, Lelani M. Mannetti, Timon McPhearson, and Tischa A. Muñoz-Erickson

Abstract We describe the rationale and framework for developing scenarios of positive urban futures. The scenario framework is conducted in participatory workshop settings and composed of three distinct scenario approaches that are used to (1) explore potential outcomes of existing planning goals (strategic scenarios), (2) articulate visions that address pressing resilience challenges (adaptive scenarios), and (3) envision radical departures from the status quo in the pursuit of sustainability and equity (transformative scenarios). A series of creative and analytical processes are used to engage the community in imagining, articulating, and scrutinizing visions and pathways of positive futures. The approach offers an alternative and complement to traditional forecasting techniques by applying inspirational stories to resilience research and practice.

Keywords Scenario development · Sustainability visioning · Anticipatory resilience · Co-production · Urban futures

---

D. M. Iwaniec (✉) · L. M. Mannetti
Andrew Young School of Policy Studies, Urban Studies Institute, Georgia State University, Atlanta, GA, USA
e-mail: diwaniec@gsu.edu

M. Berbés-Blázquez
School for the Future of Innovation in Society, Arizona State University, Tempe, AZ, USA

E. M. Cook
Environmental Science, Barnard College, New York, NY, USA

N. B. Grimm
School of Life Sciences, Arizona State University, Tempe, AZ, USA

T. McPhearson
The New School, Urban Systems Lab, New York, NY, USA

Cary Institute of Ecosystem Studies, Millbrook, New York, NY, USA

Stockholm Resilience Centre, Stockholm University, Stockholm, Sweden

T. A. Muñoz-Erickson
USDA Forest Service, International Institute of Tropical Forestry, Río Piedras, PR, USA

Z. A. Hamstead et al. (eds.), *Resilient Urban Futures*, The Urban Book Series,
https://doi.org/10.1007/978-3-030-63131-4_6

The dominant discourse about the future is dystopian. The stories told through cinema, novels, journalism, and research are full of dire warnings of catastrophes and visions of dark futures. These forecasts, predictions, and projections do provide insight. It is essential that we understand the consequences of current trends and are able to anticipate signals of impending threats. It is how we explore what to avoid and even how to survive potential collapse. However, if dystopia is the only story we tell about the future, the perceived inevitability can be a barrier to action.

Thus, we develop positive futures to realize alternatives and explore radical possibilities. In contrast to forecasts, positive futures may not be the most likely trajectory or outcome and can be rife with uncertainties. Positive futures are neither templates nor fantasies of a perfect utopia free from tradeoffs or conflict. Positive futures are stories—sometimes called scenarios—of the plausible pathways needed to achieve desirable outcomes. They are stories of possibilities to inspire and improve efforts toward achieving more sustainable, equitable, and resilient futures.

Given vast possibilities and opportunities for the future, there is a need to consider multiple, alternative future scenarios (Iwaniec et al. 2014). Neither creative nor analytical skills alone can provide the substance required for developing complex future scenarios (Wierzbicki 2007). A combination of creative and analytical skillsets is needed to craft positive futures (Wiek and Iwaniec 2014).

Visioning approaches often rely heavily on creative and unstructured processes. The creative processes lead to visions that might be inspirational, but that are not necessarily consistent, evidence-based, or plausible (Shipley 2002). At the same time, advanced visioning requires abstract reasoning, such as incorporating resilience and sustainability principles (Chap. 8). Visioning processes also require specifications to make visions tangible, for instance, by means of visualizations (Chap. 10).

In contrast to visioning, forecasting approaches can be modeled from first principles, basic assumptions, and use information about past trends and current and previous conditions to make inferences about the future. Forecasts are a suitable approach when we expect the structure of a social–ecological–technological system (SETS) to generally persist rather than fundamentally change. Positive futures, however, are intended to explore radical departures of the status quo—when small tweaks are not enough to overcome wicked problems or rapid, trend-breaking changes, and when deliberate sustainability transformations are imperative to achieve a desirable future.

A key goal of positive futures is to create space to question the limits of what is normally considered possible, desirable, or inevitable. Developing scenarios for the long term (e.g., time horizons of 50 years and longer) allows participants to navigate multiple values and explore innovative ideas for an unpredictable future. This opens up the solution-space to explore radical innovations that might require longer time horizons to unfold. Through these extended time horizons, barriers to change the current governance structure or existing infrastructure are reframed as opportunities to reimagine how urban SETS could and should work.

An emerging urban systems science, at the intersection of urban resilience, sustainability transitions, and scenario research, is beginning to focus on the crucial question of how urban SETS dynamics can be guided along more resilient, equitable, and sustainable trajectories. Cities and urban areas are complex; further, long-term futures are uncertain, subject to non-stationarity, and therefore difficult to predict and prepare for. To address this complexity, we need to challenge the dominant dystopian discourse by exploring novel, alternative, positive visions (McPhearson et al. 2017).

6.1 Approach

We describe here a framework for developing scenarios of positive urban futures. This participatory approach has been applied in nine Latin and North American cities at multiple spatial scales—neighborhood, municipal, and metropolitan region—as part of the Urban Resilience to Extreme Events Sustainability Research Network (UREx SRN; https://UREXSRN.net). Scenario development for this project focuses on articulating and exploring the implications of positive future pathways to the year 2080 for urban resilience to climate extremes (e.g., flood, drought, and heat). The emphasis on climate resilience, however, is not to limit the scope of the visions but instead to serve as a boundary condition that provides an entry point to engage in broader sustainability and resilience discussions.

6.1.1 A Framework for Positive Futures

Scenario development of positive futures is enabled by a series of steps and activities. The scenario development framework opens space to explore sustainable, resilient pathways toward SETS innovation and transformational change (Iwaniec et al. 2020a). In this chapter, we outline the sequence of key processes and activities applied in the UREx SRN project to develop multiple, alternative positive future scenarios. Further description of the rationale and methods is also provided in other chapters: analyses of past and existing vulnerabilities are presented in Chaps. 2 and 4; production of anticipatory knowledge, politics of urban resilience, and communication of climate uncertainties are explored in Chaps. 5 and 11; incorporation of existing municipal and community planning on climate resilience is described in Chap. 3; scenario co-production activities are outlined here, but further descriptions of the co-production workshop setting and approaches for stakeholder recruitment, facilitation, and addressing power dynamics are provided in Chap. 7; and evaluation and visualization approaches to explore scenario implications and assess tradeoffs are detailed in Chaps. 8–10.

The framework to develop positive futures uses three distinct scenario approaches (scenario logics; Fig. 6.1). Each of the scenario logics can be usefully applied to different contexts. Together they allow for comparative analyses among the scenarios to explore differences, evaluate tradeoffs, and build anticipatory capacity.

The key feature of this framework is the development of multiple, alternative, positive future scenarios among three distinct scenario logics.

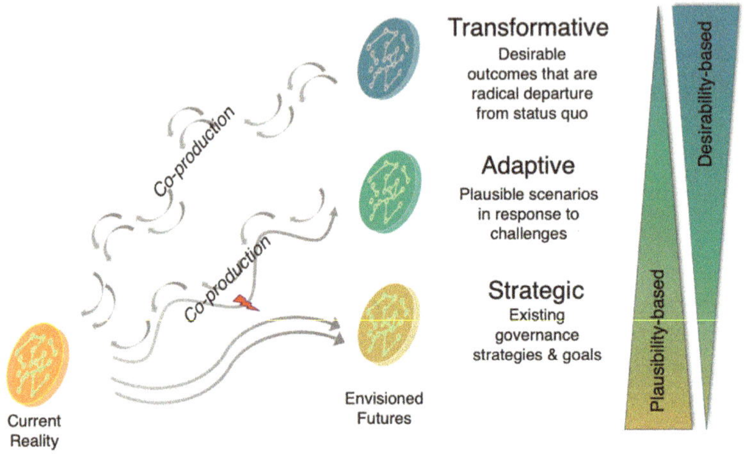

Fig. 6.1 The positive futures framework comparing strategic, adaptive, and transformative scenarios on a plausibility–desirability gradient (image modified from Iwaniec et al. 2020a). Gray arrows represent projected and backcasted scenario pathways. The red glyph indicates social, ecological, or technological disturbance(s) to address along the transition pathway. Note that the strategic scenarios are validated and explored in the workshop setting, whereas the adaptive and transformative scenarios are fully co-produced in participatory workshops

- **Strategic scenarios**are developed from existing goals and targets extracted from current plans and policies. These scenarios focus on developing long-term futures extrapolated from existing visions and plans (forecasting).
- **Adaptive scenarios** are co-produced scenarios that focus on producing social–ecological–technological innovations to address big challenges (e.g., extreme climate events). These scenarios are framed by a problem, and interventions are developed and sequenced to explore potential outcomes and tradeoffs of the scenario pathway (forecasting and backcasting).
- **Transformative scenarios**are co-produced visions and pathways that represent radical departures from the status quo in the pursuit of resilience, sustainability, and equity. These scenarios start by co-developing a vision of a desirable future and then identifying solutions and pathways linking the vision to the current state (backcasting).

All three scenario logics belong to and produce different representations of positive futures. However, there are limits to what is perceived as credible. While unexpected social changes, ecological tipping points, and technological innovations seemingly define and surprise our modern society, we generally expect current trends to persist. The scenario logics are deliberately sequenced in this framework to allow for transformative thinking—the ability to think critically about transformative change (Wolfram 2016; Iwaniec et al. 2019). The strategic, adaptive, and transformative scenarios—each with their own assumptions, concepts, translation modes, and needs for evidence-based data—vary in the production approach and the vision and pathways produced. Scenario activities are ordered such that participants first explore long-term SETS implications of their existing plans (i.e., strategic scenarios), then build on this knowledge to develop scenarios that address pressing challenges (adaptive scenarios), and scenarios that represent radical visions of sustainability, resilience, and equity (transformative scenarios).

Strategic scenarios employ an extrapolative approach to exploring long-term implications of stakeholders' existing formalized goals and targets (typically shorter-term targets, e.g., <5 years) (Iwaniec et al. 2020b). From a content analysis, strategies and actions are coded from city governance, planning, and visioning documents then clustered into distinct scenarios pathways (Chap. 3). Strategic scenarios allow participants to start from a common framing around existing formalized goals and to explore and evaluate whether these actions are sufficient to address long-term persistent and emergent challenges. These scenarios allow for exploration of the hypothesis that existing plans and policies are insufficient to address the most pressing challenges faced by cities. Participants are encouraged to understand current targets and consider the need for more ambitious solutions.

Adaptive scenarios are co-produced to explore SETS interventions that address pressing challenges (Iwaniec et al. 2020a). In this project, the adaptive scenarios focus on addressing climate change-driven extreme events (e.g., flood, drought, heat, multi-hazard disturbances). Adaptive scenarios help to build capacity for anticipatory resilience thinking (Chap. 11) through the development of novel social–ecological–technological solutions. Development of these scenarios creates space to push the boundaries of what is possible, forcing participants to ask, "Is this enough?" That is, are these futures representative of what their city should be?

Transformative scenarios are co-produced to explore radically different futures that depart from a city's current social, ecological, and technological systems (Iwaniec et al. 2020a). Although transformative scenarios among the UREx SRN cities vary greatly, they generally explore diverse and hybridized imaginaries (e.g., eco-cities, equitable cities, livable cities, self-sufficient cities, smart cities) in the context of their respective communities. The ability to think critically about transformative change can be enhanced by first (a) understanding the suite of vulnerabilities and uncertainties, (b) exploring long-term implications of existing planning goals, and (c) addressing the most pressing challenges.

6.1.2 Development of the UREx SRN Scenarios

The workshop setting brings together diverse transdisciplinary activities meant to enable detailed descriptions of scenario pathways and their constituent SETS intervention strategies. The scenario co-production process begins with a broad view needed to envision the future condition. Through iterative revision and refinement, the scenario pathways are then elaborated on to ensure coherence and tangible representations of the social-ecological-technological systems that undergo change, along with the implications of that change.

6.2 Scoping and Framing

To incorporate diverse knowledges, perspectives, and visions, scenarios may be developed in transdisciplinary and participatory settings that range from consultation to co-production (Jahn et al. 2012; Lang et al. 2012). In the case of UREx SRN, scoping and framing begins prior to the scenario development workshops. Core stakeholders help identify project-related scenario themes, potential participants, additional collaborators (see stakeholder recruitment in Chap. 7), key challenges and goals, and the temporal and spatial boundaries and scope of the scenarios.

We deploy surveys to elicit responses from a diversity of city governance actors about their perceptions of climate risks, the solutions they prefer for integration into public policy and investment decisions, how actors frame climate resilience in different contexts, and what tools and methodologies they use to collect and use climate resilience data and knowledge. The surveys also identify existing collaborations and new partnerships needed to more effectively coordinate climate resilience work across sectors. These data are used to inform workshop development and stakeholder recruitment.

In the workshop setting, we explore current vulnerabilities and projected future trends, as well as the past actions responsible for these conditions. The objective is not to create agreement at this stage but rather to identify and create a common framing around core issues. The co-production of a historical timeline of these issues is used to further build capacity for anticipatory resilience by reflecting on how the problems we face today are products of past decisions.

6.3 Goals and Intervention Strategies

In the scenario workshops, participants work in small groups to co-produce the positive future scenarios. The process is initiated by first defining the challenges and goals to be addressed in each scenario. Participants then identify initial intervention strategies needed to address these challenges and goals. Activities in this phase—such as conducting systems mapping, identifying megatrends (i.e., large, slow-moving changes) and weak signals (i.e., indicators of potentially emerging issues), and eliciting vision statements of a desirable future state—allow for rapid prototyping and brainstorming. System maps help participants refine their understanding of the relationships among the initial strategies they are considering, enabling them to brainstorm new systemic strategies and scrutinize tradeoffs among identified strategies—moreover, to produce more holistic visions. These activities are conducted both individually and as collective discussions to provide time for reflection and deliberation. Examples are provided to seed the activities with an initial pool of diverse SETS strategies and considerations. Initial seeding may represent innovations from different sources. For example, participants may look to other places facing similar challenges, scan for global megatrends that might affect their community, identify weak signals with potential for transformational change but that have not yet been scaled-up or scaled-out (Bennett et al. 2016), or inspiration from creative and fictional works. The development of radical, aspirational goals and strategies for transformative scenarios can be further facilitated by asking probing questions; for instance: Would this still be transformative/desirable in 2080? Could this be accomplished within just the next five years? What structures or power dynamics is the intervention challenging?

6.4 Scenario Specificity

Activities in this phase are designed to add details to the scenario pathway (Fig. 6.2). A key outcome is to provide enough spatial, temporal, and other key details to delineate a scenario pathway and parameterize subsequent modeling and assessments (Table 6.1).

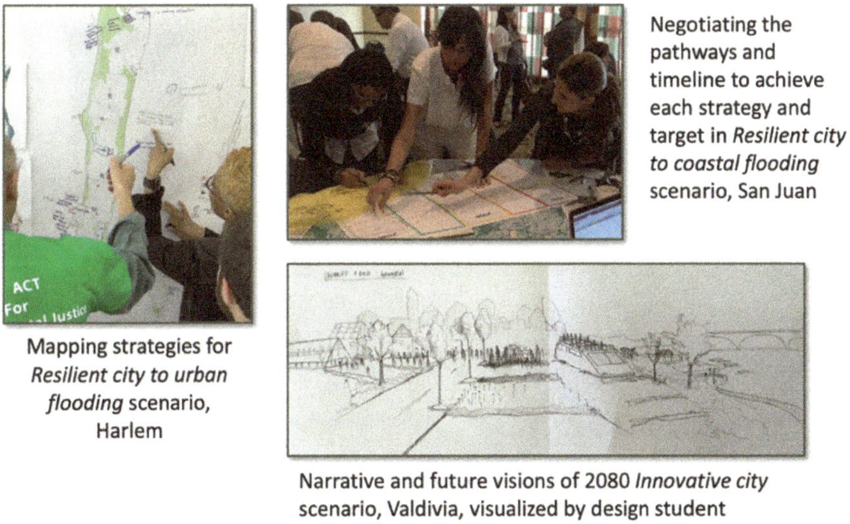

Fig. 6.2 Photos from scenario workshops of participants engaging in spatial and temporal specificity activities

Table 6.1 Description of activities carried out in the scenario specificity phase

Activity	Description	Example
Target specificity	Defining targets, indicators, and metrics for the strategies	A shade infrastructure strategy might describe a target of 30% canopy cover of native trees and 10% cover with solar energy-producing shade structures—this may also include indicators and metrics for achieving targets associated with heat mitigation, biodiversity, and access to greenspace.
Spatial specificity	GIS-based participatory mapping of the locations and other spatial characteristics (e.g., size, configuration, amount of centralization/decentralization) that identify where and for whom particular strategies will be implemented	Participants draw the specific locations, size, and configuration of new greenspace on a map. However, since not all strategies can be easily represented this way; they are also articulated as rules (e.g., trees are to be sited along auxiliary streets in the poorest neighborhoods).
Temporal specificity	Sequencing the strategies as a scenario pathway along a timeline—new SETS strategies are often added during this activity to detail what is needed to enable or support the intended changes or how to maintain it once implemented	Details are provided for when the tree planting initiative starts (e.g., 2020), the rate of implementation and corresponding intermediate targets, and when the implementation of the strategy is to be completed (e.g., 2045).
Governance specificity	Identifying key actors and institutions responsible for implementing each strategy	New partnerships and institutions, and their roles and actions may be detailed. More transformative examples describe radical reconfiguration of power regimes, new governance structures, changes in culture, and empowered communities.
Normative specificity	Describing the multiple value-laden objectives and implications of the envisioned strategies	A "day in the life" narrative describes what a future person in 2080 experiences and how they interact with the vision.

6.5 Evaluation and Dissemination

The scenarios co-produced through subsequent steps of the workshop represent future visions and the pathways to reach them. The outcome is a diverse suite of alternative, plausible visions. These scenarios may be represented and evaluated through qualitative assessments (Chap. 8), quantitative modeling (Chap. 9), design-based

Fig. 6.3 Design-based renderings of positive future scenarios from Phoenix, USA (for more details on these scenarios, see https://sustainability.asu.edu/future-scenarios/)

renderings (Fig. 6.3), and data visualization tools (Chap. 10) to explore potential implications and compare tradeoffs of the alternative visions. Through iterative evaluation of the diverse outcomes, the scenario pathways can be refined to better reflect desired outcomes.

Dissemination of positive futures entails more than just meaningfully conveying the final products. Dissemination activities should occur throughout the process so as to open dialogue and involve a broader community to further elicit diverse preferences, check representativeness, develop opportunities for further engagement, and support transparency. Chapter 10 describes dissemination approaches to democratize decision-making through data visualization tools. Implementation programs can vary greatly in scope. They may focus on incorporating the goals, strategies, and targets from this work into formal planning documents. Alternatively, an implementation program may focus on more bottom-up processes, such as supporting existing community initiatives that align with the transition pathway or developing champions for new initiatives that serve as a key leverage point for change. Chapter 11 concludes with a discussion on embedding future scenarios into current planning practices toward resilient urban futures.

6.6 Conclusion

The approach described in this chapter offers various options for co-producing scenarios of positive futures that may be applied across organizational units (e.g., municipal, private, non-profit institutions), sectors (e.g., housing, energy, water, transportation, food, health), spatial scales (from global to local), and temporal scales (10 to >100 years into the future). Although the activities described in this chapter may not all be appropriate in all contexts, the activities are meant to be flexible and crafted to match the varied needs, available resources, capacities, and objectives of the project. For example, project objectives may include overcoming conflicts between long-term ambitions and short-term concerns. Such conflicts call for additional emphasis on developing anticipatory and long-term thinking capacities. Similarly, an objective of the project may include the need to address conflicts among divergent city and community preferences or priorities. To address this divergence, the workshops may require broader engagement and activities that further emphasize the development of normative capacities.

A critical tension in positive futures is the need for evidence-based representations of an envisioned future and the ability to portray radical transformations of novel conditions. Various approaches exist to bridge these needs, such as descriptive, empirical, and modeling work from other places, as well as the application of concept proofs, pilot projects, and experimentation. Moreover, a portfolio of diverse scenarios can also help address this tension. Multiple alternative scenarios of different types and degrees of change can be used in a participatory setting to evaluate what is "too radical" and what is "not transformative enough" to achieve a desired future state.

Fundamentally, the process of co-producing scenarios of positive futures can help to build anticipatory, long-term, normative, and systems thinking capacities. With these key capacities, cities can increase their agency to successfully implement future resilience, sustainability, and transformational change initiatives (Romero-Lankao et al. 2016; Wolfram 2016; Iwaniec et al. 2019). However, regardless of how motivating the positive futures are, alone, they are generally not sufficient for catalyzing transformational change; they need to be incorporated into dissemination and implementation programs. In some cases, "windows of opportunities" can arise when a positive vision of a sustainability transition has been developed prior to a disaster, and its uptake is enabled by the disaster (Birkmann et al. 2010; Brundiers and Eakin 2018; Solecki et al. 2019). The central goal of this emerging urban systems science, however, should be to guide and facilitate anticipatory change without loss to human well-being, ecological integrity, and critical infrastructure.

The framework for positives futures integrates three distinct scenario logics into a structured transdisciplinary research-practice approach to develop future scenarios. The development of strategic, adaptive, and transformative scenarios is used to envision innovative solutions and interventions, contrast plausible-desirable visions and pathways of the future, address future challenges, realize opportunities, and explore radical possibilities.

References

Bennett EM, Solan M, Biggs R et al (2016) Bright spots: seeds of a good Anthropocene. Front Ecol Environ 14:441–448. https://doi.org/10.1002/fee.1309

Birkmann J, Buckle P, Jaeger J, Pelling M, Setiadi N, Garschagen M, Fernando N, Kropp J (2010) Extreme events and disasters: a window of opportunity for change? Analysis of organizational, institutional and political changes, formal and informal responses after mega-disasters. Nat Hazards 55:637–655. https://doi.org/10.1007/s11069-008-9319-2

Brundiers K, Eakin HC (2018) Leveraging post-disaster windows of opportunities for change towards sustainability: a framework. Sustainability 10:1390. https://doi.org/10.3390/su10051390

Iwaniec DM, Childers DL, VanLehn K, Wiek A (2014) Studying, teaching and applying sustainability visions using systems modeling. Sustainability 6:4452–4469. https://doi.org/10.3390/su6074452

Iwaniec DM, Cook EM, Barbosa O, Grimm NB (2019) The framing of urban sustainability transformations. Sustainability 11:573. https://doi.org/10.3390/su11030573

Iwaniec DM, Cook EM, M. Davidson et al (2020a) The co-production of sustainable future scenarios. Landscape and Urban Plan 197:103744.https://doi.org/10.1016/j.landurbplan.2020.103744

Iwaniec DM, Davidson MJ, Cook EM et al (2020b) Integrating existing climate adaptation planning into future visions: a strategic scenario for the central Arizona-Phoenix region. Landscape Urban Plan 200:103820. https://doi.org/10.1016/j.landurbplan.2020.103820

Jahn T, Bergmann M, Keil F (2012) Transdisciplinarity: between mainstreaming and marginalization. Ecol Econ 79:1–10. https://doi.org/10.1016/j.ecolecon.2012.04.017

Lang DJ, Wiek A, Bergmann M et al (2012) Transdisciplinary research in sustainability science: practice, principles, and challenges. Sustain Sci 7:25–43. https://doi.org/10.1007/s11625-011-0149-x

McPhearson T, Iwaniec DM, Bai X (2017) Positive visions for guiding urban transformations toward sustainable futures. Curr Opin Environ Sustain 22:33–40. https://doi.org/10.1016/j.cosust.2017.04.004

Romero-Lankao P, Gnatz DM, Wilhelmi O, Hayden M (2016) Urban sustainability and resilience: from theory to practice. Sustainability 8:1224. https://doi.org/10.3390/su8121224

Shipley R (2002) Visioning in planning: Is the practice based on sound theory? Environ Plan A 34:7–22. https://doi.org/10.1068/a3461

Solecki W, Grimm NB, Marcotullio P, Boone CG, Bruns A, Lobo J, Lueke AE, Romero-Lankao P, Young AF, Zimmerman R, Breitzer R, Giffith C, Aylett A (2019) Extreme events and climate adaptation-mitigation linkages: Understanding low-carbon transitions in the era of global urbanization. Wires Climate Change 10(6):e616. https://doi.org/10.1002/wcc.616

Wiek A, Iwaniec DM (2014) Quality criteria for visions and visioning in sustainability science. Sustain Sci 9:497–512. https://doi.org/10.1007/s11625-013-0208-6

Wolfram M (2016) Conceptualizing urban transformative capacity: A framework for research and policy. Cities 51:121–130. https://doi.org/10.1016/j.cities.2015.11.011

Wierzbicki AP (2007) Modelling as a way of organising knowledge. Eur J Oper Res 176:610–635. https://doi.org/10.1016/j.ejor.2005.08.018

David M. Iwaniec is an Assistant Professor of Urban Sustainability at the Urban Studies Institute, Andrew Young School of Policy Studies at Georgia State University. He is a sustainability scientist researching anticipatory and systems approaches to advance urban sustainability, resilience, and justice. His work focuses on the co-development of scenarios and transition pathways for positive futures of urban transformation

Marta Berbés-Blázquez is an Assistant Professor at the School for the Future of Innovation at Arizona State University. Her research considers the human dimensions of social-ecological transformations in rural and urban ecosystems with an emphasis on vulnerable populations. Her work is informed by resilience thinking and political ecology at a conceptual level, and it is practically oriented toward qualitative, participatory, and anticipatory research methods. Specific topics of expertise include power dynamics and access in ecosystem services, scenario planning, resource extraction, eco-health, climate change adaptation, and transformation.

Elizabeth M. Cook is an Assistant Professor at Barnard College in the Department of Environmental Science. She is an urban ecosystem ecologist and her research focuses on future urban sustainability and human-environment feedbacks in urban and nearby native ecosystems. She conducts research on sustainability and resilience planning through participatory scenario development with local stakeholders. Her work seeks to understand cities as social-ecological-technological systems with a comparative approach in Latin American and U.S. cities. Cook holds a Ph.D. from Arizona State University and a Bachelor of Arts from Wellesley College.

Nancy B. Grimm studies the interactions between climate variation and change, human activities, and ecosystems. Her interdisciplinary urban and stream ecosystem research focuses on disturbance, resilience, and biogeochemical processes. Dr. Grimm is Regents Professor at Arizona State University and currently co-directs the Urban Resilience to Extremes Sustainability Research Network (UREx SRN). She was president (2005-06) and is a fellow of the Ecological Society of America, as well as a fellow of the American Geophysical Union (AGU), the American Association for the Advancement of Science (AAAS), and the Society for Freshwater Science. She is an elected member (2019) of the US National Academy of Sciences.

Lelani M. Mannetti is a Postdoctoral Research Associate at the Urban Studies Institute, conducting comparative research across nine US and Latin American cities as part of the Urban Resilience to Extreme Events Sustainability Research Network. Alongside city stakeholders, she helps promote visionary thinking through the co-production of desirable and plausible future scenarios focused on resilience, sustainability, and equity. Dr. Mannetti is also a Fellow on the Intergovernmental Science-Policy Platform on Biodiversity and Ecosystem Services Values Assessment, a methodological assessment of the diverse conceptualization of multiple values of nature and its benefits.

Timon McPhearson is Director of the Urban Systems Lab and Associate Professor of Urban Ecology at The New School. He is a Research Fellow at The Cary Institute of Ecosystem Studies and Stockholm Resilience Centre and a member of the IPCC.

Tischa A. Muñoz-Erickson is a Research Social Scientist in the USDA Forest Service's International Institute of Tropical Forestry, in Río Piedras, Puerto Rico. She studies urban sustainability governance, including the policy networks, knowledge systems, anticipatory capacities, and strategies to advance sustainability, resilience, and equity. She is also actively involved in transdisciplinary platforms to facilitate the co-production of futures and transition pathways in the US and Latin American cities. Muñoz-Erickson leads the International Urban Field Station and the San Juan ULTRA and is also Co-PI of the National Science Foundation Urban Resili-ence to Extreme Events Sustainability Research Network (UREx SRN) and the SETS Convergence project.

Chapter 7
Setting the Stage for Co-Production

Elizabeth M. Cook, Marta Berbés-Blázquez, Lelani M. Mannetti, Nancy B. Grimm, David M. Iwaniec, and Tischa A. Muñoz-Erickson

Abstract Participatory scenario visioning aims to expose, integrate, and reconcile perspectives and expectations about a sustainable, resilient future from a variety of actors and stakeholders. This chapter considers the settings in which transdisciplinary participatory visioning takes place, highlighting lessons learned from the Urban Resilience to Extremes Sustainability Research Network (UREx SRN). It reflects on the benefits of engaging in the co-production process and the challenges that must be considered amid this process.

Keywords Co-production · Participatory · Decision-making · Governance

7.1 Co-Production to Address Urban Resilience Challenges

Envisioning positive change can help urban leaders imagine and transition to more sustainable and resilient futures for cities. Cities face seemingly insurmountable and complex resilience challenges—supporting transparent and just governance systems, reducing environmental inequities, addressing failing infrastructure—all compounded by the uncertainties of climate change (Elmqvist 2018; Rosenzweig

E. M. Cook (✉)
Department of Environmental Science, Barnard College, New York City, NY, USA
e-mail: ecook@barnard.edu

M. Berbés-Blázquez
School for the Future of Innovation in Society, Arizona State University, Tempe, AZ, USA

L. M. Mannetti · D. M. Iwaniec
Andrew Young School of Policy Studies, Urban Studies Institute, Georgia State University, Atlanta, GA, USA

N. B. Grimm
School of Life Sciences, Arizona State University, Tempe, AZ, USA

Julie Ann Wrigley Global Institute of Sustainability, Arizona State University, Tempe, AZ, USA

T. A. Muñoz-Erickson
USDA Forest Service, International Institute of Tropical Forestry, Rio Piedras, Puerto Rico, USA

Z. A. Hamstead et al. (eds.), *Resilient Urban Futures*, The Urban Book Series,
https://doi.org/10.1007/978-3-030-63131-4_7

and Solecki 2018; van der Heijden 2019). Collectively, we need to consider not just what can or could happen, but also what *ought* to happen to ensure a city is resilient in the face of climate change.

Imagining outside-of-the-box ideas in alternative future visions can inspire innovative solutions to meet city goals (Bai et al. 2016; Bennett et al. 2016; McPhearson et al. 2016; Pereira et al. 2018; Iwaniec et al. 2020; Chap. 6). In a complex system such as a city, envisioning the future must incorporate values and expertise from diverse communities and sectors. As a form of participatory engagement, we focus on transdisciplinary co-production of knowledge and solutions that reflect the diverse values, knowledge, and future expectations of the stakeholders involved.

Participatory processes can take many forms, and the engagement setting differs depending on the project goals, resources and capacities, and disciplinary perspectives (Miller and Wyborn 2018; Wyborn et al. 2019). In planning, participatory processes engage residents directly in decisions about their communities, exemplified by processes such as participatory budgeting. In this book and chapter, we refer to participatory to mean the engagement of diverse stakeholders, inclusive of decision-makers, community leaders, community members, or academics. Here, the participatory process is the active involvement of academic and non-academic stakeholders in sharing ideas or providing feedback through workshops, interviews, focus groups, or surveys. Co-production, on the other hand, more specifically refers to co-developing or co-learning new ideas or forms of knowledge together. Co-production of ideas goes beyond just sharing ideas or extracting information from a particular group; it often involves reconciling differences and finding new, shared understandings through an interactive and iterative participatory process.

In sustainability science, co-production is an instrumental approach to resilience and sustainability planning, or more generally to problem-solving, that involves active, collaborative engagement with diverse partners. This approach is often time-consuming and messy. However, we posit that a pluralistic co-production of ideas enhances our ability to advance future urban planning with more sustainable solutions for urban transformation (Pereira et al. 2018; Iwaniec et al. 2019; Elmqvist et al. 2019). Moreover, engaging an inclusive set of partners is key to developing legitimate, actionable, and salient research and policy agendas (Schwarz and Herrmann 2016; Acuto et al. 2018; Wyborn et al. 2019; Norström et al. 2020; Ruiz-Mallén 2020).

In this chapter, we briefly describe the co-production process in the Urban Resilience to Extremes Sustainability Research Network (UREx SRN). To inform the development of co-production visioning projects, we build on our collective experience from the UREx SRN, as well as existing literature, by highlighting some lessons learned on key elements of meaningful co-production and the challenges that can arise during the process.

7.2 Co-Production of Positive Long-Term Visions in the UREx SRN

In the UREx SRN, the co-production approach was designed to guide a participatory process of long-term scenario visioning to address urban resilience and sustainability challenges in nine cities, each with their own unique partners and goals. The co-production process began in each city with an initial scoping phase. The scoping phase identified potential partners, needs of the city, preliminary shared goals, and key themes for the future visions. It was completed through three steps: (1) one-on-one discussions or world-café (i.e., round robin) idea generation with academic and non-academic partners, (2) a governance document analysis to capture the existing goals and strategies contained in city plans (see Chap. 3), and (3) a governance survey to capture a broader set of actors' visions and expectations beyond the formal, governmental (dominant) visions (see Chap. 6; Muñoz-Erickson et al. *under review*). The visioning process was intended as an intervention or catalyst in support of governance processes and dynamics in which the cities were already immersed. For example, in 2015 in Valdivia, Chile, a Sustainability Plan of Action (City of Valdivia 2015) was developed in collaboration with the InterAmerican Development Bank; the UREx SRN co-production process built upon the existing Sustainability Plan to further develop the goals with specific targets and to challenge the visions toward being even more ambitious, creative, and transformative.

The co-production of future visions occurred primarily through a participatory workshop setting. The workshops were designed to bring together different sectors and actors to deliberate and work toward a shared articulation of plausible and desirable futures—positive futures. The participatory workshops centered around developing alternative visions during facilitated small-group and plenary activities with approximately 35–45 participants (see Chap. 6). The scenarios formed the basis for framing the diverse challenges cities face, understanding and exploring feedbacks and tradeoffs of future decisions, as well as guiding pathways for future decision-making.

7.3 Elements of Co-Production

Drawing from our collective experiences and a significant body of literature, we highlight four key elements that have been critical in the UREx SRN approach to co-production: focusing on process, finding a collective commitment, ensuring credibility and legitimacy of the work, and capturing a diversity of perspectives (Fig. 7.1; Box 7.1).

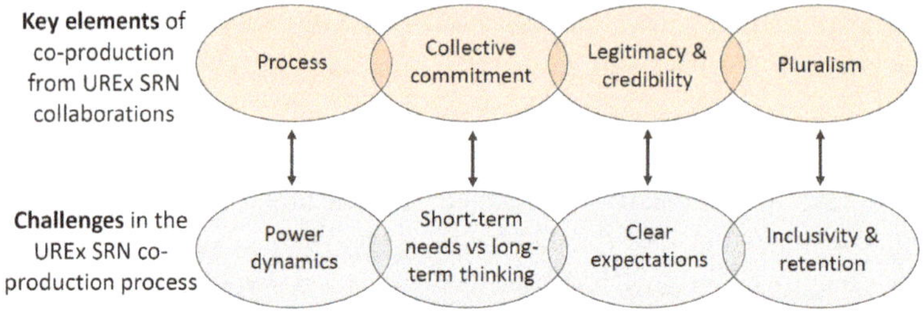

Fig. 7.1 Key elements and challenges that can arise in the participatory process of co-production

7.3.1 Process and Outcomes

Co-production centers process as much as it centers outcomes. Meaningful outcomes can include the emergence of new and shared knowledge, innovative solutions with targets, new relationships, or the development of new collective capacities. Achieving these goals requires an iterative process with flexible short- and long-term plans and a mix of creative forms of engagement (Cvitanovic et al. 2019). As a collaborative and inclusive endeavor, co-production includes academic and non-academic stakeholders as full equal partners, each contributing diverse expertise. It places stakeholders' knowledge, opinions, and aspirations at the center of the process. In practice, this requires active engagement and commitment throughout the project, including the initial framing and budgeting to equitably account for individuals' contributions.

7.3.2 Collective Commitment

Partners in meaningful co-production need a collective commitment—a shared goal to address the challenges of a particular place (Norström et al. 2020). The collective commitment can be determined in the initial scoping phase. The commitment must reflect an acknowledgement of the history and politics involved in that place and a willingness to listen, share, and learn new ideas as the project evolves. The process inevitably will involve challenges, such as reconciling contested meanings, exploring new uncertainties, and navigating power dynamics and dissent. However, by creating a deliberate space for co-learning around a collective goal, these problems can be addressed. The facilitators must make space for building rapport, trust, and meaningful relationships through an iterative, reflective, and flexible process. For example, through activities focused on unraveling assumptions and exploring new ideas, participants may find commonalities in which they reconcile contested meaning and build a shared—sometimes new—understanding of future urban resilience challenges (Galafassi et al. 2018).

7.3.3 Credibility and Legitimacy

The co-production process can produce actionable and credible knowledge for urban planning (Nevens et al. 2013; Muñoz-Erickson et al. 2017). It acknowledges that integrating diverse stakeholder perspectives has the essential role of making evident the complexities and needs of the system. Thus, drawing together the local knowledge of participants, co-production can improve capacity for developing credible strategies and solutions that account for the nuances of the city. Through the co-production process, participants describe all aspects of these solutions: the goals and targets of the interventions and who will benefit from them; the underlying values, knowledge, evidence, and potential tradeoffs; the envisioned actions, plus the needed roles and responsibilities to achieve those actions; and who is involved in implementation and governance. Thus, this collaborative process increases the legitimacy and credibility of shared solutions that can be championed by both communities and decision-makers.

7.3.4 Diversity of Perspectives

Co-production is a pluralistic process (Norström et al. 2020). The process can create new roles for partners, equalize power dynamics, and prioritize a more equitable representation of diverse perspectives. For example, participatory processes can be an opportunity for marginalized voices in the community to lead in idea generation and planning, thus contributing to future policy decisions. The process integrates scientific and non-scientific inquiry and various forms of knowledge—including those from impacted communities—blurring the distinction between the knowledge producer and the knowledge user (Muñoz-Erickson 2014). In addition, new networks are created in which marginalized communities are equal partners with—or in some instances, given preference over—the dominant voices and visions driving decisions on future urban resilience. Participation in the co-production process can also build multiple capacities, including adaptive (Eakin et al. 2014), anticipatory (see Chaps. 11 and 6), and transformative capacities (Wolfram 2016; Wolfram et al. 2019; Chap. 6).

Box 7.1 Operationalizing co-productionin the UREx SRN

The **process and outcomes** of practitioner–researcher interactions in Hermosillo, Mexico: Through an iterative process, the Hermosillo city stakeholders and research partners engaged equally in the scenario development process through group discussions (virtual and in-person), joint research, and collective workshop design. The diverse forms of knowledge and ideas were reflected in both themes and goals of the ultimate scenario visions, which ranged from addressing urban safety to extreme heat stress. After participatory visioning workshops, the transdisciplinary team formed thematic working groups, *Mesas de Trabajo*, to continue integrating the co-developed visions into upcoming governance plans. ***Outcomes***: new partnerships, shared knowledge, continued engagement in *Mesas de Trabajo* beyond the participatory visioning workshop, contribution to planning documents.

Collective commitment from the practitioner-research team in Valdivia, Chile: A strong rapport and collective commitment existed between the city stakeholders and researchers, who shared an understanding of the city's challenges related to wetland conservation and flooding. Despite challenges and uncertainties throughout the process (e.g., funding, changing roles and jobs, election of new officials with different priorities, and evolving needs), an integrated future vision was co-produced through on-going engagement resulting from meaningful, trust-based relationships. Throughout, new ideas were openly explored and a common framework for urban sustainability, resilience, and transformations was developed and reconciled as needed. ***Outcomes:*** integrated vision with concrete strategies.

Enhancing **credibility and legitimacy** of governance frameworks for resilience in Portland, Oregon: Meaningful collaborations were formed with the Disaster Resilience and Recovery Action Group, the City of Portland, and UREx SRN that built on earlier participatory scenario efforts (Resilience Infrastructure Planning Exercise, 2018). The co-production process uncovered the complexities of the city's governance structures and the subtle, often-hidden barriers that needed to be overcome. The collaborations resulted in a shift in mindset from "resilience-as-harm-reduction" to "resilience-as-thriving" before and after a disaster. By synthesizing perspectives and finding areas of strategic overlap among different visioning participants, the credibility of shared strategies and goals was enhanced. ***Outcomes***: emergent governance structures and shared principles for resilience.

Empowering **diverse perspectives** by co-producing future scenarios in Phoenix, Arizona: The series of scenario workshops in Phoenix was attended by a diverse group of community leaders and decision-makers, including representatives from federal, tribal, state, county, and city agencies, as well as non-governmental organizations and universities. Here, and across UREx SRN cities, creative and analytical activities captured the diverse ways in which desirable futures are envisioned among participants in the workshops. Co-production activities included different forms of scoping and framing (described above), structured knowledge-sharing conversations, World Cafés with a round-robin format to build ideas, individual and collective brainstorming, creative storytelling, and spatial and temporal participatory mapping. ***Outcomes***: integration of different forms of scientific and non-scientific analysis and knowledge systems, including those from marginalized communities.

7.4 Confronting the Challenges of Co-Production

Co-production is touted as a way to improve decision-making and research on urban resilience; however, it entails important challenges, such as power dynamics, a tension between short- and long-term thinking, unclear expectations, and failure to include all relevant perspectives (Fig. 7.1; Box 7.2; Lemos et al. 2018; Wyborn et al. 2019; Jagannathan et al. 2020). Dealing with these challenges—which evolve throughout a project—requires flexibility, reflexivity, and open communication.

7.4.1 Power Dynamics

Power differentials underpin the interactions between groups that have patently diverse and competing agendas, priorities, assumptions, and ways of understanding the city. Uneven power dynamics surface while engaging with diverse stakeholders (Johnson et al. 2016; Frantzeskaki and Rok 2018; Turnhout et al. 2020). Although experienced facilitators and careful activity design can help to navigate power dynamics, they never fully go away and thus must be acknowledged and managed. Likewise, some degree of tension can be productive; a workshop where everyone agrees probably signals that disagreement is simply repressed, in which pre-existing power dynamics are "containing" new or radical ideas and forcing superficial consensus. This repression can lead to lack of engagement, distrust, and ultimately a loss of project legitimacy. Too much tension is equally detrimental. Walking this fine line is an art and a science, but facilitators can help by being transparent about goals and intentions, establishing and respecting boundaries, sitting with the tensions (rather than offering solutions), creating an atmosphere where people feel safe (but not necessarily comfortable), and centering and amplifying underrepresented voices. The organizing team should take care to structure activities such that all voices are equally valued and scrutinized. Moreover, the activities must work toward building bridges and weaving existing knowledge into something new, rather than forcing integration that may lead to privileging one form of knowledge over another.

7.4.2 Short-Term Needs and Long-Term Thinking

The UREx SRN workshops are often situated in a context that requires balancing long-term visioning with short-term needs. Addressing future resilience and climate challenges requires a long-term perspective paired with a long-term commitment to the (often slow) co-production process. The co-production visioning process requires that communities have the time, resources, and capacity to focus on long-term future visioning, alongside more urgent and immediate needs (Berbés-Blázquez et al. 2018; Turnhout et al. 2020). The long-term future can seem intractable or even irrelevant in the face of current needs and challenges of communities vulnerable to flooding, heat stress, or other climate resilience challenges. Action is needed now. At the same time, investing in short-term solutions triggered by crisis events can result in maladaptation (Anderson et al. 2018). To facilitate this discussion, it is helpful to articulate the need for anticipatory, long-term futures in order to make better decisions today, as well as to reflect on the fact that problems of today are the product of past decisions. Ideally, co-production processes go beyond one-time initiatives toward a sustained process where both short- and long-term strategies are routinely explored, stress-tested, and evaluated to allow for exploration of uncertainties and generation of creative works.

7.4.3 Clear Expectations

There are often disconnects among the needs and priorities of partners from academia, communities, and civic and municipal organizations. For both academic and non-academic partners, it is critical to set clear expectations—acknowledging the limitations of the project, the scope of resources, and intended outcomes (Jagannathan et al. 2020). Expectations should be tailored based on the resources available to bridge existing, on-going work in cities with the co-production of long-term goals and visions. Likewise, clear guidelines should be set to ensure the products and follow-up are timely, accessible, and useful for partners. The value of products that can arise from the same work may vary greatly. For example, academic outputs in peer-reviewed journals, for which researchers will receive credit in their institutions, are often inaccessible to non-academic partners and may not highlight the most relevant outcomes or actionable pathways that practitioners need in implementing outcomes. There is, therefore, a need to develop a communication strategy that includes non-academic products, such as reports, websites, guides and manuals, or podcasts that speak to broader audiences. The project must set aside resources and time for continuous outreach and dissemination of findings with the team.

7.4.4 Inclusivity and Retention

Inclusivity and retention of diverse perspectives in the process—not just the most vocal or dominant views—is essential. It can be difficult to determine who to involve in the co-production process (Frantzeskaki and Rok 2018). Some projects may focus on expert-led discussions while others will feature community-led discussions or a diverse mix; regardless, the process will gain legitimacy with a balanced representation of appropriate partners, each open to disparate world views. Retention can also be a challenge, particularly as the co-production process and research outputs are typically slow. Co-production necessitates flexibility in daily obligations and work expectations; participating is a privilege of those with time, resources, and flexibility

(Frantzeskaki and Rok 2018; Cvitanovic et al. 2019). Likewise, participants' attention and time may be drawn to more pressing needs. For non-academic stakeholders, this may require resources, such as monetary compensation for transportation and child care, or approval by employers for time spent participating in the activities outside of daily responsibilities. On the other hand, academics may need to press for institutional support that acknowledges the time, effort, and relevance of participatory research and engagement with non-academic stakeholders.

Box 7.2 Confronting co-productionchallenges in the UREx SRN

Addressing conflict and power dynamics: Conflicts will arise and are a signal of meaningful co-production through hard conversations. Yet, power dynamics must be managed to allow for equitable contributions. In the UREx SRN co-production process, small-group facilitators participate in pre-workshop training to help address power dynamics among participants, and activities are centered on negotiation and deliberate consensus building. Similarly, throughout the workshop, participants are asked to reflect on a set of "ground rules" agreed upon at the beginning, such as considering how much they are speaking and if they need to "step up or step back" in their role at the table.

Reconciling short-term needs and long-term thinking: Long-term, positive visioning is a useful tool to think beyond the current system constraints and to avoid focusing on small tweaks. However, current needs and short-term implementation plans must also be addressed. In Baltimore, Maryland, to center long-term visions addressing multi-jurisdictional watershed management on near-term transitions, the team met in a series of follow-up workshops to develop actionable implementation (5–10 years) timelines. The timelines highlighted short-term and specific metrics, budgeting, financing, and governance mechanisms, new partnerships, and communication and education strategies. In other cities, such as San Juan, Puerto Rico, and Hermosillo, Mexico, practitioner-research teams developed *Mesas de Trabajo* (working groups) to further address short-term implementation plans.

Establishing clear expectations: Setting expectations about timeline, commitment, resources, and anticipated outcomes early in the process is critical. With nine cities, a challenge in the UREx SRN co-production process has been maintaining meaningful engagement and producing context-dependent outputs on a timescale relevant to practitioner and community needs in all cities. Through a Knowledge-Action Taskforce, we work with city teams to offer a series of products, including the Future Cities podcast (in Spanish and English), the interactive Urban Resilience data visualization platform (https://urex.urbansystemslab.com/), story maps, reports and slides highlighting modeling and scenario outputs, as well as academic papers.

Maintaining inclusivity and retention: In UREx SRN visioning workshops, participants varied throughout the process and across the cities. For example, in Phoenix, Arizona and New York City, New York, emphasis was placed on including local planning officials and community leaders, whereas in Miami, Florida and Syracuse, New York, participants comprised academics and city practitioners focused on resilience and recovery. Awareness of the ways in which local contexts, governance structures, and knowledge networks can affect participation is critical in keeping participants engaged and collectively working toward the desired outcomes. Across a network of nine diverse cities, the challenge of keeping participants engaged is significant and requires considerably more time, institutional and financial support, expertise, personal commitment to build meaningful relationships, as well as flexibility, creativity, diplomacy, and patience. A risk lies in opting for familiar partners and approaches, and to this end, regular UREx network evaluations have proven useful.

7.5 Moving Co-Production Forward

Co-production presents important benefits and opportunities for urban transformation and resilience planning. There is value in creating a space for creativity, interacting with new individuals, and co-learning. Yet, co-production is at times aspirational in its goals and the challenges of this process must be acknowledged and addressed. When careful attention is paid to the process of collaborative engagement, the co-production process has potential to build capacity for on-going co-learning and continued engagement in resilience planning.

References

Acuto M, Parnell S, Seto KC (2018) Building a global urban science. Nat Sustain 1(1):2. https://doi.org/10.1038/s41893-017-0013-9

Anderson SE, Bart RR, Kennedy MC et al (2018) The dangers of disaster-driven responses to climate change . Nat Clim Change 8(8):651–653. https://doi.org/10.1038/s41558-018-0208-8

Bai X, van der Leeuw S, O'Brien K et al (2016) Plausible and desirable futures in the Anthropocene: a new research agenda. Global Environ Change 39:351–362. https://doi.org/10.1016/j.gloenvcha.2015.09.017

Bennett EM, Solan M, Biggs R et al (2016) Bright spots: seeds of a good Anthropocene. Front Ecol Environ 14(8):441–448. https://doi.org/10.1002/fee.1309

Berbés-Blázquez M, Iwaniec D, Grimm N et al (2018) Positive visions for sustainable, resilient, and equitable cities. The Nature of Cities. https://www.thenatureofcities.com/2018/04/21/positive-visions-sustainable-resilient-equitable-cities/. Accessed 15 Jun 2020

City of Valdivia (2015) Plan de Acción de Valdivia, Chile. Programa Ciudades Emergentes y Sostenibles (CES). https://webimages.iadb.org/PDF/PLAN+DE+VALDIVIA+27-05.pdf. Accessed 07 Jul 2020

Cvitanovic C, Howden M, Colvin RM et al (2019) Maximising the benefits of participatory climate adaptation research by understanding and managing the associated challenges and risks. Environ Sci Policy 94:20–31. https://doi.org/10.1016/j.envsci.2018.12.028

Eakin HC, Lemos MC, Nelson DR (2014) Differentiating capacities as a means to sustainable climate change adaptation. Global Environ Change 27:1–8. https://doi.org/10.1016/j.gloenvcha.2014.04.013

Elmqvist T (2018) The urban planet: knowledge towards sustainable cities. Cambridge University Press, Cambridge, UK

Elmqvist T, Andersson E, Frantzeskaki N et al (2019) Sustainability and resilience for transformation in the urban century. Nat Sustain 2(4):267–273. https://doi.org/10.1038/s41893-019-0250-1

Frantzeskaki N, Rok A (2018) Co-producing urban sustainability transitions knowledge with community, policy and science. Environ Innov Soc Trans 29:47–51. https://doi.org/10.1016/j.eist.2018.08.001

Galafassi D, Daw TM, Thyresson M et al (2018) Stories in social-ecological knowledge cocreation. Ecol Soc 23(1):23. https://doi.org/10.5751/ES-09932-230123

Iwaniec DM, Cook EM, Barbosa O et al (2019) The framing of urban sustainability transformations. Sustainability 11(3):573. https://doi.org/10.3390/su11030573

Iwaniec DM, Cook EM, Davidson MJ et al (2020) The co-production of sustainable future scenarios. Landscape Urban Plann 197:103744. https://doi.org/10.1016/j.landurbplan.2020.103744

Jagannathan K, Arnott JC, Wyborn C et al (2020) Great expectations? reconciling the aspiration, outcome, and possibility of co-production . Curr Opin Environ Sustain 42:22–29. https://doi.org/10.1016/j.cosust.2019.11.010

Johnson JT, Howitt R, Cajete G et al (2016) Weaving indigenous and sustainability sciences to diversify our methods. Sustain Sci 11:1–11. https://doi.org/10.1007/s11625-015-0349-x

Lemos MC, Arnott JC, Ardoin NM et al (2018) To co-produce or not to co-produce. Nat Sustain 1(12):722–724. https://doi.org/10.1038/s41893-018-0191-0

McPhearson T, Iwaniec DM, Bai X (2016) Positive visions for guiding urban transformations toward sustainable futures. Curr Opin Environ Sustainability 22:33–40. https://doi.org/10.1016/j.cosust.2017.04.004

Miller CA, Wyborn C (2018) Co-production in global sustainability: histories and theories. Environ Sci Policy. https://doi.org/10.1016/j.envsci.2018.01.016

Muñoz-Erickson TA (2014) Co-production of knowledge–action systems in urban sustainable governance: the KASA approach. Environ Sci Policy 37:182–191. https://doi.org/10.1016/j.envsci.2013.09.014

Muñoz-Erickson TA, Miller CA, Miller TR (2017) How cities think: knowledge co-production for urban sustainability and resilience . Forests 8(6):203. https://doi.org/10.3390/f8060203

Nevens F, Frantzeskaki N, Gorissen L et al (2013) Urban transition labs: co-creating transformative action for sustainable cities. J Cleaner Prod 50(1):111–122. https://doi.org/10.1016/j.jclepro.2012.12.001

Norström AV, Cvitanovic C, Löf MF et al (2020) Principles for knowledge co-production in sustainability research. Nat Sustain 1–9. https://doi.org/https://doi.org/10.1038/s41893-019-0448-2

Pereira LM, Hichert T, Hamann M et al (2018) Using futures methods to create transformative spaces: visions of a good Anthropocene in southern Africa. Ecol Soc 23(1):19. https://doi.org/10.5751/ES-09907-230119

Rosenzweig C, Solecki W (2018) Action pathways for transforming cities. Nat Clim Change 8:756–759. https://doi.org/10.1038/s41558-018-0267-x

Ruiz-Mallén I (2020) Co-production and resilient cities to climate change. In: Nared J, Bole D (eds) Participatory research and planning in practice. Springer International Publishing, Cham, Switzerland, pp 1–11

Schwarz K, Herrmann DL (2016) The subtle, yet radical, shift to ecology for cities. Front Ecol Environ 14(6):296–297. https://doi.org/10.1002/fee.1288

Turnhout E, Metze T, Wyborn C et al (2020) The politics of co-production XE "co-production": participation, power, and transformation. Curr Opin Environ Sustain 42:15–21. https://doi.org/10.1016/j.cosust.2019.11.009
van der Heijden J (2019) Studying urban climate governance: where to begin, what to look for, and how to make a meaningful contribution to scholarship and practice. Earth Syst Gover 1:100005. https://doi.org/10.1016/j.esg.2019.100005
Wolfram M (2016) Conceptualizing urban transformative capacity: a framework for research and policy. Cities 51:121–130. https://doi.org/10.1016/j.cities.2015.11.011
Wolfram M, Borgström S, Farrelly M (2019) Urban transformative capacity: from concept to practice. Ambio 48(5):437–448. https://doi.org/10.1007/s13280-019-01169-y
Wyborn C, Datta A, Montana J et al (2019) Co-producing sustainability: reordering the governance of science, policy, and practice. Annu Rev Environ Resour 44:319–346. https://doi.org/10.1146/annurev-environ-101718-033103

Elizabeth M. Cook is an Assistant Professor at Barnard College in the Department of Environmental Science. She is an urban ecosystem ecologist and her research focuses on future urban sustainability and human-environment feedbacks in urban and nearby native ecosystems. She conducts research on sustainability and resilience planning through participatory scenario development with local stakeholders. Her work seeks to understand cities as social-ecological-technological systems with a comparative approach in Latin American and U.S. cities.

Marta Berbés-Blázquez is an Assistant Professor at the School for the Future of Innovation at Arizona State University. Her research considers the human dimensions of social-ecological transformations in rural and urban ecosystems with an emphasis on vulnerable populations. Her work is informed by resilience thinking and political ecology at a conceptual level, and it is practically oriented toward qualitative, participatory, and anticipatory research methods. Specific topics of expertise include power dynamics and access in ecosystem services, scenario planning, resource extraction, eco-health, climate change adaptation, and transformation.

Lelani M. Mannetti is a Postdoctoral Research Associate at the Urban Studies Institute at Georgia State University. She conducts comparative research across nine US and Latin American cities as part of the Urban Resilience to Extreme Events Sustainability Research Network. Alongside city stakeholders, she helps promote visionary thinking through the co-production of desirable and plausible future scenarios focused on resilience, sustainability, and equity. Dr. Mannetti is also a Fellow on the Intergovernmental Science-Policy Platform on Biodiversity and Ecosystem Services Values Assessment, a methodological assessment of the diverse conceptualization of multiple values of nature and its benefits.

Nancy B. Grimm studies the interactions between climate variation and change, human activities, and ecosystems. Her interdisciplinary urban and stream ecosystem research focuses on disturbance, resilience, and bio-geochemical processes. Dr. Grimm is Regents Professor at Arizona State University and currently co-directs the Urban Resilience to Extremes Sustainability Research Network (UREx SRN). She was president (2005-06) and is a fellow of the Ecological Society of America, as well as a fellow of the American Geophysical Union (AGU), the American Association for the Advancement of Science (AAAS), and the Society for Freshwater Science. She is an elected member (2019) of the US National Academy of Sciences.

David M. Iwaniec is an Assistant Professor of Urban Sustainability at the Urban Studies Institute, Andrew Young School of Policy Studies at Georgia State University. He is a sustainability scientist researching anticipatory and systems approaches to advance urban sustainability, resilience, and justice. His work focuses on the co-development of scenarios and transition pathways for positive futures of urban transformation.

Tischa A. Muñoz-Erickson is a Research Social Scientist in the USDA Forest Service's International Institute of Tropical Forestry, in Río Piedras, Puerto Rico. She studies urban sustainability governance, including the policy networks, knowledge systems, anticipatory capacities, and strategies to advance sustainability, resilience, and equity. She is also actively involved in transdisciplinary platforms to facilitate the co-production of futures and transition pathways in the US and Latin American cities. Muñoz-Erickson leads the International Urban Field Station and the San Juan ULTRA and is also Co-PI of the National Science Foundation Urban Resilience to Extreme Events Sustainability Research Network (UREx SRN) and the SETS Convergence project.

Chapter 8
Assessing Future Resilience, Equity, and Sustainability in Scenario Planning

Marta Berbés-Blázquez, Nancy B. Grimm, Elizabeth M. Cook, David M. Iwaniec, Tischa A. Muñoz-Erickson, Vivian Hobbins, and Darin Wahl

Abstract In the absence of strong international agreements, many municipal governments are leading efforts to build resilience to climate change in general and to extreme weather events in particular. However, it is notoriously difficult to guide and activate processes of change in complex adaptive systems such as cities. Participatory scenario planning with city professionals and members of civil society provides an opportunity to coproduce positive visions of the future. Yet, not all visions are created equal. In this chapter, we introduce the Resilience, Equity, and Sustainability Qualitative (RESQ) assessment tool that we have applied to compare positive scenario visions for cities in the USA and Latin America. We use the tool to examine the visions of the two desert cities in the Urban Resilience to Extreme Events Sustainability Research Network (UREx SRN), which are Hermosillo (Mexico) and Phoenix (United States).

M. Berbés-Blázquez (✉)
School for the Future of Innovation in Society, Arizona State University, Arizona, USA
e-mail: mberbes@asu.edu

N. B. Grimm
School of Life Sciences, Arizona State University, Arizona, USA

Julie Ann Wrigley Global Institute of Sustainability, Arizona State University, Tempe, AZ, USA

E. M. Cook
Environmental Science, Barnard College, New York, NY, USA

D. M. Iwaniec
Andrew Young School of Policy Studies, Urban Studies Institute, Georgia State University, Georgia, USA

T. A. Muñoz-Erickson
USDA Forest Service, International Institute of Tropical Forestry, Río Piedras, PR, USA

V. Hobbins
Instituto Tecnológico de Sonora, Sonora, Mexico, USA

D. Wahl
Lund University Centre for Sustainability Studies, Lund, Sweden

Z. A. Hamstead et al. (eds.), *Resilient Urban Futures*, The Urban Book Series,
https://doi.org/10.1007/978-3-030-63131-4_8

Keywords Extreme weather events · RESQ assessment · Climate resilience · Sustainability · Future visions

Scenario planning is a tool to inform management actions in situations of deep uncertainty. It is therefore not surprising that in recent decades, scenario approaches have been used to explore complex issues such as climate change adaptation, urbanization, or biodiversity conservation. The scenario process produces a set of alternative futures of a place, or a situation, that allow comparisons between the outcomes of adopting different policy decisions. In doing so, scenarios not only offer a way of representing complexity but also a means for comparison. The ability to explicitly compare the positive and negative outcomes of competing policy options makes scenarios a valuable instrument to support and guide decision-making.

There are different ways of assessing and exploring the desirability of alternative scenario visions. Traditionally, scenario work has relied on modeling outputs as a means of comparison. Models can show quantitative differences in key variables of the system, such as water use or land use, which are easy to grasp and relate directly to policy goals. At the same time, models have limitations. The most obvious is that modeling is restricted to the aspects of the future visions that have a quantifiable, biophysical expression. Therefore, intangible qualities that might be highly desirable, such as creating a sense of place or valuing the history of a neighborhood, cannot be captured by models. This means that one of the main strengths of the scenario technique, which is the production of rich, textured, nuanced depictions of the future based on the integration of different ways of knowing, is left out of the evaluative part of the exercise because of the lack of a physical approximation. In this chapter, we introduce the Resilience, Equity, and Sustainability Qualitative (RESQ) assessment tool that we have applied to compare coproduced scenarios for cities in the USA and Latin America. We use as examples two heat and drought scenarios that were developed in the Sonoran Desert cities of Hermosillo (Mexico) and Phoenix (USA). This pair of scenarios illustrate how two cities facing the same climate challenge might envision their future differently. In our work, the RESQ assessment is complementary to other comparative analyses, including models for future land use/land change (see Chap. 7), rich narratives, and visual renderings of the future. Beyond evaluation, having a variety of synthetic outputs from scenario planning offers a range of entry points with which to engage a wider audience. The chapter is structured as follows: we first provide definitions for the key concepts of resilience, equity, and sustainability before we delve into the case studies for Hermosillo and Phoenix. We then reflect on the potential of the RESQ assessment as a heuristic tool for social learning, and briefly explore its strengths and limitations.

8.1 An Instrument for Assessment

8.1.1 Defining Resilience, Equity, Sustainability

A broad goal of the UREx SRN project is to create positive urban visions to assist cities in building resilience to extreme events, advancing sustainability, and ensuring equity. All of these terms are used widely and inconsistently in the literature. Hence, in this section, we present our definitions for resilience, sustainability, and equity and explain our approach to their assessment.

8.1.1.1 Resilience

Resilience, as used in the context of social–ecological–technological systems, refers to the capacity of a complex adaptive system to absorb disturbance while remaining within a given domain of attraction that is defined by structure, function, identity, and feedbacks (Carpenter et al. 2001). Climate resilience, which is relevant to our work, adds specificity to Carpenter's definition and is understood as the ability of communities to persist, adapt, and recover in the face of climate stresses and shocks (Tanner et al. 2009; Wardekker et al. 2010). Increasingly, there have been calls to provide resilience metrics to help guide management decisions (Quinlan et al. 2015). While it is possible to identify concrete metrics for the resilience of well-studied systems like lakes—where nutrient loading is the key variable that determines whether the lake is in its eutrophic or oligotrophic state—it is a great deal more difficult to identify and measure metrics for complex and contested systems like cities (Meerow et al. 2016). Doing so is further complicated by the fact that our visions represent future states, and hence, indicators that might be relevant today might not be relevant in the future, nor would they capture the idiosyncrasies of cities around the world. Hence, our assessment focuses on the identification and evaluation of general mechanisms that enhance resilience and are applicable to a variety of complex systems. Our list of principles is based on Walker and Salt (2006), Chapin et al. (2009), and Biggs et al. (2012), and is as follows:

- **Diversity.** Actions that increase or maintain diversity increase the resilience of a social–ecological–technological system. Diversity can mean increasing variety (how many distinct elements there are), balance (how many of each distinct element there are), or disparity (the extent to which elements are distinct from one another). Strategies to increase diversity can be ecological, such as introducing pollinator gardens in front lawns; social, such as opening decision-making processes to a broader set of actors; and technical, such as amplifying the kinds of energy sources that make up a city's energy portfolio.
- **Functional redundancy.** Resilience is enhanced by mechanisms that build redundancy, that is, by having components within the system that fulfill similar or overlapping functions that will allow the system to continue to operate even if a

specific component fails. For instance, natural and mechanical infrastructures can be combined to create shade in public spaces. Both provide cooling benefits, but they are affected by extreme events differently; for example, a drought can kill vegetation but will not affect shade infrastructure.

- **Connectivity.** Managing the links between different components of a system is important for resilience. Increasing connectivity is necessary at times, as when people come together after a disaster; at other times, decreasing connectivity is needed; for example, quarantining a diseased population.
- **Slow variables.** The resilience of a system is largely dependent on critical variables that control its internal dynamics and generally change at a gradual pace (see Walker et al. 2012). For example, land-use zoning determines density in cities, which is a slow variable that in turn may promote or inhibit services such as mass transit, social programs, or green belts.
- **Feedbacks.** The resilience of a system depends on reinforcing or dampening responses in reaction to an initial stimulus (positive or negative feedback loops), which may occur at a later time or in a distant location. Sometimes feedbacks are set up intentionally; for example, using pricing mechanisms to manage water usage. Other times, feedbacks are manifestations of unintended consequences; for example, adding road lanes to reduce gridlock usually has the opposite effect.
- **Complex system lens.** Adopting a holistic perspective that seeks to understand adaptive behavior builds resilience when dealing with uncertainty. A systems lens considers not only the social, ecological, and technological elements of the system but also their interactions across scales. Extreme events often demonstrate the interconnectedness of systems. The aftermath of Hurricane Maria in Puerto Rico in 2017 exemplified the degree to which vulnerability is a product of social–ecological–technological flows of information, financial transactions, people, and materials—all of which need to be considered simultaneously and across scales (Eakin et al. 2018).
- **Learning.** Learning is a key in situations of uncertainty and can take many forms, from awareness-raising campaigns to changes in the educational curriculum. We pay special attention to processes leading to double-loop learning (*sensu* Argyris 1999) and social learning (Reed et al. 2010), which have the potential to transform societal values.
- **Adaptive management.** The practice of adaptive environmental management is used under conditions of uncertainty. It is often dubbed "learning by doing" because it intends to improve our understanding of how a system works through management policies that are taken to be hypotheses. Therefore, adaptive management encourages experimentation, monitoring, and iterative decision-making.
- **Participation.** Broadening participation means that decisions are made based on a more complete understanding of the issue built on a variety of perspectives. This leads to better-informed decisions, with more buy-in and empowerment of participants.
- **Polycentric governance.** In polycentric governance, there are several nodes with decision-making capacity that act with some degree of autonomy from one another, and there is usually an effort to match the governing authority with the

scale of the problem. For example, in some jurisdictions, there may be water boards that make decisions for a watershed, working in conjunction with one another but separate from other branches of government.

8.1.1.2 Sustainability and Equity

The term sustainability was first coined in 1972 by the Club of Rome to seek a balance between satisfying the needs of people and respecting the biophysical limits of the planet's ecosystems. Thus, sustainability sought to temper the push for economic development with a long term, ecological perspective. It was understood that sustainability was critical to guide development in the Global South. Fifteen years later, the World Commission on Environment and Development published *Our Common Future,* where they defined sustainable development as "meet[ing] the needs of the present without compromising the ability of future generations to meet their own needs" (Brundtland 1987, p. 43). Hence, the Brundtland Report made intra and intergenerational equity considerations more central to the definition. The Brundtland Report also established the idea of sustainability in the international policy arena. Over the years, global initiatives have sought to operationalize sustainability into the development agendas of individual countries. One such initiative was the Sustainable Development Goals (SDGs), which were adopted at the 2012 United Nations Conference on Sustainable Development in Rio de Janeiro. The SDGs replaced the earlier iteration of global targets for sustainable development, the Millennium Development Goals (2000). Compared with the Millennium Development Goals, the SDGs are more comprehensive in scope, including goals like job creation, peace building, and responsible consumption; as such, the SDGs are broadly applicable to both rich and poor nations. The 17 SDGs are as follows:

- **Goal 1.** End poverty in all its forms everywhere.
- **Goal 2.** End hunger, achieve food security and improved nutrition, and promote sustainable agriculture.
- **Goal 3**. Ensure healthy lives and promote well-being for all at all ages.
- **Goal 4.** Ensure inclusive and equitable quality education and promote life-long learning opportunities for all.
- **Goal 5.** Achieve gender equality and empower all women and girls.
- **Goal 6.** Ensure availability and sustainable management of water and sanitation for all.
- **Goal 7.** Ensure access to affordable, reliable, sustainable, and modern energy for all.
- **Goal 8.** Promote sustained, inclusive and sustainable economic growth, full and productive employment and decent work for all.
- **Goal 9.** Build resilient infrastructure, promote inclusive and sustainable industrialization, and foster innovation.
- **Goal 10.** Reduce inequality within and among countries.
- **Goal 11.** Make cities and human settlements inclusive, safe, resilient and sustainable.

- **Goal 12**. Ensure sustainable consumption and production patterns.
- **Goal 13.** Take urgent action to combat climate change and its impacts.
- **Goal 14.** Conserve and sustainably use the oceans, seas and marine resources for sustainable development.
- **Goal 15.** Protect, restore and promote sustainable use of terrestrial ecosystems, sustainably manage forests, combat desertification, and halt and reverse land degradation and halt biodiversity loss.
- **Goal 16.** Promote peaceful and inclusive societies for sustainable development, provide access to justice for all and build effective, accountable and inclusive institutions at all levels.
- **Goal 17.** Strengthen the means of implementation and revitalize the global partnership for sustainable development.

We use the SDGs to assess equity and sustainability in our scenario visions. However, we have modified this list in two ways in our assessment: First, we consider the subgoals associated with SDG11, since it pertains to cities. Hence, we are assessing explicitly the following subgoals: (1) providing safe and affordable housing; (2) having access to safe, affordable, and sustainable transportation; (3) participatory governance for urban planning; (4) protecting the world's cultural and natural heritage; (5) reducing the adverse environmental impact of cities (e.g., cutting atmospheric pollution); (6) access to green and public spaces; (7) considering multiscalar relations between urban, per-urban, and rural areas; and (8) building sustainable and resilient buildings with local materials. Second, we have excluded SDG 13 (taking action on climate change) because it is repetitive, as well as SDG 17 (revitalizing the global partnership for sustainable development) because it is not applicable at the scale of our city scenarios.

8.1.2 Qualitative Assessment—How It Works

An important part of the RESQ tool is to maintain a systems view of the future vision. Thus, the assessment begins by identifying the three to five defining characteristics of the scenario that capture its identity. These constitute the vision's key components sensu Holling's (2001) "rule of hand," which states that even very complex systems can be understood by identifying three to five key interacting components that organize all the rest. Thus, the rule of hand organizes the variables of the system in a hierarchy to reduce the amount of complexity and avoid being bogged down by details. This is especially useful when dealing with data from scenario processes that produce rich and nuanced depictions of the future. In our work, we use the rule of hand to characterize the essence of a vision co-developed in scenario workshops, where participants construct visions through guided activities that include brainstorming, goal-setting, mapping exercises, prioritization, and story-telling (see Iwaniec et al. 2020). The actual process for determining the key components is

Fig. 8.1 Principles associated with resilience, equity, and sustainability used in our qualitative assessment

similar to qualitative coding, in which underlying themes are identified by carefully studying data from the workshops, including notes, maps, and narratives. For example, increasing green infrastructure might emerge as a key component of a future vision, and subsumed within it might be several more detailed strategies, such as building bioswales, providing subsidies for green roofs in commercial buildings, targeting educational campaigns for school children, and updating the building code for new construction to require green features. In order to ensure consistency in our work, at least three researchers reviewed and agreed on the identification of key components.

After the key components are identified, they are assessed against the resilience principles and the SDGs goals identified earlier (Fig. 8.1) using a Likert scale from 0 (absent) to 4 (present). Again, this assessment is qualitative and considers all of the outputs from the workshop activities. In the green infrastructure example, the key component of increasing green infrastructure would get high scores on principles related to greening the city (SDG 11.7) and lower scores on principles related to polycentric governance. Because future visions are complex, the key components that make up a vision may have very different foci that build different aspects of resilience, equity, and/or sustainability. Hence, we use an average of the scores assessed for each key component when presenting our results. Each key component was scored by two researchers separately and reviewed by a third.

8.2 Comparing Drought and Heat Scenarios

Cities around the world will face more intense and frequent climate-related extreme events in the future. In this section, we consider how two arid-land cities envision coping with or adapting to more extreme heat, longer drought periods, and water shortages. Both Hermosillo and Phoenix are situated in the Sonoran Desert, approximately 500 km apart as the crow flies. Hermosillo is the capital city of the state of Sonora, with a population just shy of one million people. Phoenix is the capital of the state of Arizona, with a metro population of nearly 5 million people. Climate models for the Sonoran Desert region predict an increase in temperature and a decline in overall precipitation, but with particular regard to winter-spring precipitation that generates most of the water supply, as well as to the decline of snowpack at higher elevations that supports flow in major water-supply rivers (Wuebbles et al. 2017). This combination of higher temperatures with lower winter precipitation and snowpack means that droughts will last longer and be more frequent. It is clear from this projection that water security is a pressing issue for both municipalities.

The two visions showcased as examples in this chapter were created following the scenario method outlined in Chapter 6 (see also Iwaniec et al. 2020). In this form of scenario planning, workshops are used to convene a group of 30–40 municipal practitioners, representatives of civil society, and academics. The group is divided into thematic work tables, each with five or six participants. The themes of the work tables are decided in advance but defined broadly enough to accommodate diverse viewpoints in the workshop. The themes become scenarios, or visions, through the activities of the workshop. Typical themes of the UREx SRN scenarios explored the adaptation to extreme events, such as heat, drought, and flooding. We refer to these as adaptive scenarios. There were also transformative scenarios that considered the future of broader, usually normative, issues such as equity, transportation, or housing. Throughout the day participants worked with a facilitator on different activities designed to identify large goals and specific strategies associated with those goals, each defined in terms of time (when they will occur) and space (where they will be implemented). Although we produced five visions in Phoenix and six visions in Hermosillo, for simplicity, we are choosing to compare the adaptive scenarios that had to do with building resilience to heat and drought in each city.

8.2.1 Identifying Key Components

The scenario vision for Hermosillo was built around three key components, the first being aggressive water conservation measures. These measures ranged from developing wastewater treatment and reuse infrastructure, to installing rainwater harvesting devices in households, to promoting xeric landscaping on public lands. The second key component emphasized the establishment of natural areas for water-source protection and to increase green infrastructure. For example, participants envisioned reducing agricultural land use and setting aside 50,000 hectares of forest in the watershed to protect the headwaters, and creating a network of green corridors and filtration ponds to reduce the amount of flash floods. The third key component of the Hermosillo vision aimed to raise awareness and capacity to manage water use through a variety of strategies that included general education campaigns, apps to monitor water quality and quantity, education materials for schools, and the creation of a network of multidisciplinary experts to weigh in on water management decisions.

The Phoenix vision also had three key components. Similar to Hermosillo, the first key component emphasized drastic water conservation measures with similar strategies around water harvesting and wastewater reuse; however, in Phoenix, participants also envisioned centralized, underground water banking as a key piece of their vision to ensure the city's future water supply. The next key component for the Phoenix scenario was directed at managing urban density, thus, there was a push to limit sprawl, increase density in the city core, and reduce periurban agriculture, which is currently a large water consumer. To make up for the loss of agricultural productivity, the third key component of this vision aimed to increase urban agriculture through the proliferation of community gardens, vertical gardens of drought-tolerant plants, and the growth of farm-to-table movements.

8.2.2 Assessing Resilience-Building Mechanisms

The resilience assessments for the Hermosillo and Phoenix visions are shown in Fig. 8.2. The assessment for the Hermosillo vision shows that it had well-developed mechanisms for increasing redundancy, connectivity, diversity, and participation. Redundancy and diversity tend to score similarly because often, strategies that increase redundancy also increase diversity. In this case, the vision showed that participants not only identified a broad variety of strategies for reducing water consumption, but that these strategies had functional overlap. The Hermosillo vision also scored high in participation as it contained many strategies for capacity-building and public engagement; for example, involving residents in water monitoring and education campaigns for different audiences. Finally, and related to participation, the strategies that increased connectivity in the scenario were mostly social and had to do with efforts to connect and inform people about water issues.

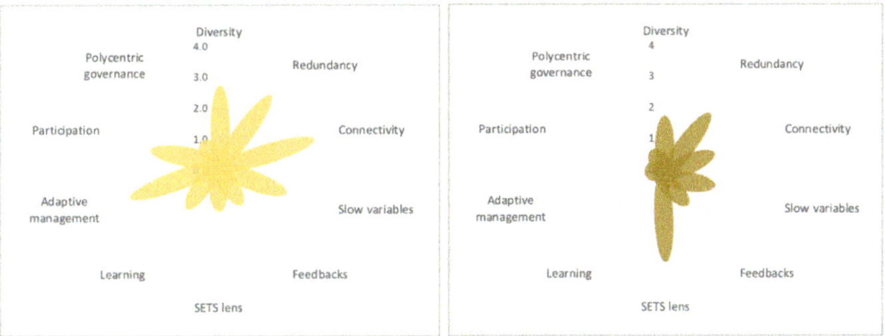

Fig. 8.2 Comparison of resilience-building mechanisms in the drought and heat scenario visions for Hermosillo (left) and Phoenix (right)

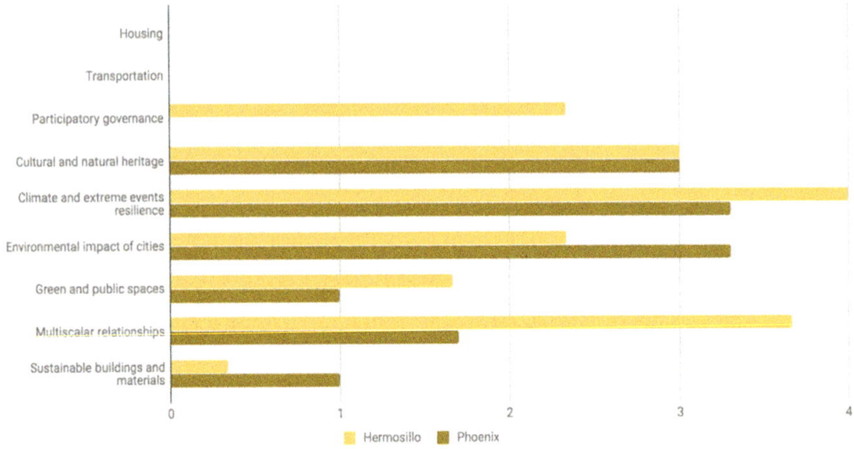

Fig. 8.3 Comparison of subgoals associated with Sustainable Development Goal 11: Sustainable Cities and Communities, in Hermosillo and Phoenix

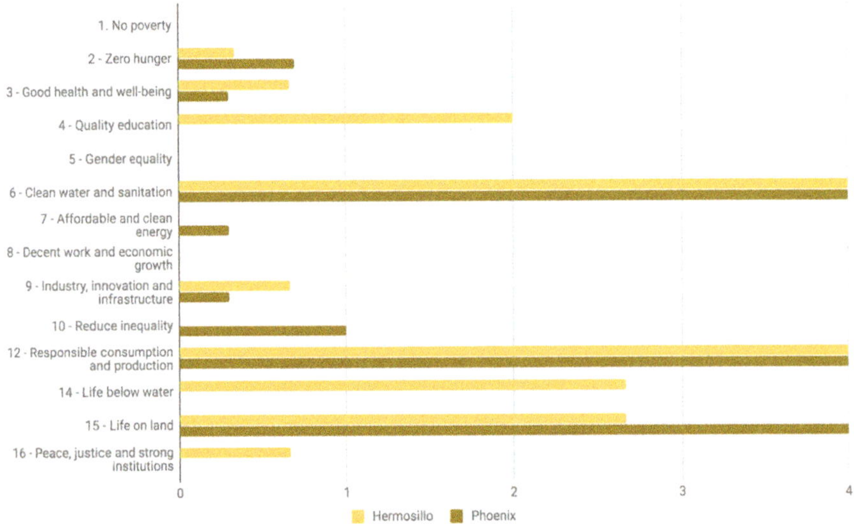

Fig. 8.4 Comparison of Sustainable Development Goals for Hermosillo and Phoenix

On the other hand, the vision of Phoenix scored highest in terms of adopting a systems lens. A systems lens means that there is evidence that participants were considering the interactions among social, ecological, and technological elements across scales. For instance, the way in which participants managed density and land-use change in the Phoenix vision showed that they were considering the links between urban and periurban regions, as well as social, ecological, and technological repercussions of reducing agriculture and increasing urban density. The Phoenix vision also scored high on redundancy since it had many overlapping strategies for water conservation happening at different scales, from high-efficiency plumbing in households to centralized water banking (see Fig. 8.2).

When we consider the two visions in general, we notice that both visions scored higher on the principles that are pertinent to changing parts of the system structure (i.e., increasing diversity, redundancy, and connectivity). Both visions had average scores on the principles that encourage systemic thinking (i.e., considering slow variables, feedbacks, and using systems thinking lens), and they scored lowest, particularly Phoenix, on the principles related to management and governance (i.e., practicing adaptive management, learning, participation, and polycentricity). When we compare the two assessments, we note that the drought vision for Hermosillo has more developed resilience-building mechanisms than the vision created for Phoenix, with the biggest differences being the degree of participation and the wish to implement adaptive management. This may reflect the cities' differences in the degree to which drought and water shortage are considered to be challenges today: Hermosillo has more of a history of water scarcity, including water rationing (Eakin et al. 2007), than does Phoenix.

8.2.3 Assessing Sustainability and Equity

The two scenarios scored low in the assessment of SDG11′s subgoals and the sustainable development goals in general (Figs. 8.3 and 8.4). This likely reflects the fact that both of these visions represent adaptive scenarios, which are issue-driven scenarios, instead of transformative scenarios, which emphasize a more holistic vision for an improved city (see Chap. 6). What this means is that participants focused more narrowly on creating a vision to solve the drought issue and did not consider other aspects that make cities livable. Topics like housing, transportation, gender equity, poverty alleviation, or job creation seem to have been absent from the conversation, despite important linkages between them and drought. On the other hand, both scenarios scored high on the SDG goals that were related to water quality and quantity and the impacts of climate change.

8.3 Discussion and Conclusion

In sustainability science, scenarios have long been hailed as a tool for exploring the normative dimensions of alternative futures pertinent to policy-making (Swart et al. 2004; Wiek and Iwaniec 2014, Raudsepp-Hearne et al. 2019). Yet, the degree to which scenarios influence actual policy decisions remains to be seen. In terms of framing issues, scenarios have found more use during initial stages of the policy cycle than in policy design and implementation (Volkery and Ribeiro 2009). Beyond institutional barriers, one outstanding issue is that normative evaluation is inherently difficult because it involves assessing values and beliefs. We see the RESQ assessment as a launching point to open a dialog on these normative dimensions of the future.

The main strength of the RESQ is that it serves as a heuristic tool for comparison and reflection. For example, in some of the UREx SRN project cities, we held a subsequent workshop where we presented the RESQ assessment to the participants who created the visions and prompted them to reconsider what worked and what was missing from their ideal future. It is through this reflection that participants realize how their vision may have resulted in a wealth of diversification strategies yet had left out consideration of governance mechanisms. The RESQ is most useful when used comparatively. For example, stakeholder groups that created different visions might compare their RESQ scores and get ideas on strategies and mechanisms to adapt for their own vision. It is in this sense that the RESQ assessment supports processes of coproduction and social learning—the sustained learning that happens through continued interactions and deliberation that change people's attitudes and behaviors (Reed et al. 2010).

At the same time, decision-makers often want to see metrics that can be used to monitor progress and impacts of a policy choice. Although we have steered clear of creating an index, the RESQ assessment can be a first step toward developing more concrete indicators that are relevant to the local context. For instance, if a city vision emphasizes mechanisms to manage slow variables, there could be a number of social, ecological, and technological indicators that can be developed, such as surveying changes in people's attitudes over time, checking groundwater recharge rates, or monitoring land-use change in the city. These indicators should be developed in discussion with participants to ensure relevancy to their context.

Finally, a warning, any assessment tool that tries to capture the nature of a complex adaptive system is grappling with incommensurability (Quinlan et al. 2015). Even a qualitative assessment such as the RESQ risks oversimplification. Thus, we offer this tool as a means to aid the policy process and not as an end in itself (Stirling 1999). That is, the process of making decisions about how a society should enhance resilience, equity, and sustainability is a great deal more political and contested than the RESQ, or any other tool for assessment, suggests.

References

Argyris C (1999) On organizational learning, 2nd edn. Blackwell, Oxford

Biggs R, Schlüter M, Biggs D et al (2012) Toward principles for enhancing the resilience of ecosystem services. Annu Rev Environ Resour 37:421–448

Brundtland G (1987) Report of the World Commission on Environment and Development: Our Common Future. United Nations General Assembly. Available via https://sustainabledevelopment.un.org/milestones/wced. Accessed 17 June 2020.

Carpenter S, Walker B, Anderies JM, Abel N (2001) From metaphor to measurement: resilience of what to what? Ecosystems 4(8):765–781

Chapin FS III, Kofinas GP, Folke C et al (2009) Principles of ecosystem stewardship: resilience-based natural resource management in a changing world. Springer-Verlag, New York

Eakin H, Magaña V, Smith J et al (2007) A stakeholder driven process to reduce vulnerability to climate change in Hermosillo, Sonora, Mexico. Mitigat Adapt Strateg Glob Chan 12(5):935–955
Eakin H, Muñoz-Erickson TA, Lemos MC (2018) Critical lines of action for vulnerability and resilience research and practice: lessons from the 2017 hurricane season. J Extreme Events 5(2):1850015
Holling CS (2001) Understanding the complexity of economic, ecological, and social systems. Ecosystems 4(5):390–405
Iwaniec DM, Cook EM, Davidson M et al (2020) The co-production of sustainable future scenarios. Landscape Urban Plann 197:103744
Meerow S, Newell JP, Stults M (2016) Defining urban resilience: a review. Landscape Urban Plann 147:38–49
Quinlan AE, Berbés-Blázquez M, Haider LJ et al (2015) Measuring and assessing resilience: broadening understanding through multiple disciplinary perspectives. J Appl Ecol 53(3):677–687
Raudsepp-Hearne C, Peterson GD, Bennett EM et al (2019) Seeds of good anthropocenes: developing sustainability scenarios for Northern Europe. Sustain Sci 15(2):605–617
Reed M, Evely AC, Cundill G et al (2010) What is social learning? Ecol Soc 15(4)
Stirling A (1999) The appraisal of sustainability: some problems and possible responses. Local Environ 4(2):111–135
Swart RJ, Raskin P, Robinson J (2004) The problem of the future: sustainability science and scenario analysis. Glob Environ Chan 14(2):137–146
Tanner T, Mitchell T, Polack E et al (2009) Urban governance for adaptation: assessing climate change resilience in ten Asian cities. Institute of Development Studies Working Papers, pp 1–47
Volkery A, Ribeiro T (2009) Scenario planning in public policy: understanding use, impacts and the role of institutional context factors. Technol Forecast Soc Chan 76(9):1198–1207
Walker B, Carpenter S, Rockstrom J et al (2012) Drivers, "slow" variables, "fast" variables, shocks, and resilience. EcolSoc 17(3)
Wardekker JA, de Jong A, Knoop JM et al (2010) Operationalising a resilience approach to adapting an urban delta to uncertain climate changes. Technol Forecast Soc Chan 77(6):987–998
WiekA ID (2014) Quality criteria for visions and visioning in sustainability science. Sustain Sci 9(4):497–512
Wuebbles DJ, Fahey DW, Hibbard KA et al (eds) (2017) Climate science special report: fourth national climate assessment, vol I. U.S, Global Change Research Program, Washington, DC

Marta Berbés-Blázquez is an Assistant Professor at the School for the Future of Innovation at Arizona State University. Her research considers the human dimensions of social-ecological transformations in rural and urban ecosystems with an emphasis on vulnerable populations. Her work is informed by resilience thinking and political ecology at a conceptual level, and it is practically oriented toward qualitative, participatory, and anticipatory research methods. Specific topics of expertise include power dynamics and access in ecosystem services, scenario planning, resource extraction, eco-health, climate change adaptation, and transformation.

Nancy B. Grimm studies the interactions between climate variation and change, human activities, and ecosystems. Her interdisciplinary urban and stream ecosystem research focuses on disturbance, resilience, and bio-geochemical processes. Dr. Grimm is Regents Professor at Arizona State University and currently co-directs the Urban Resilience to Extremes Sustainability Research Network (UREx SRN). She was president (2005-06) and is a fellow of the Ecological Society of America, as well as a fellow of the American Geophysical Union (AGU), the American Association for the Advancement of Science (AAAS), and the Society for Freshwater Science. She is an elected member (2019) of the US National Academy of Sciences.

Elizabeth M. Cook is an Assistant Professor at Barnard College in the Department of Environmental Science. She is an urban ecosystem ecologist and her research focuses on future urban sustainability and human-environment feedbacks in urban and nearby native ecosystems. She conducts research on sustainability and resilience planning through participatory scenario development with local stakeholders. Her work seeks to understand cities as social-ecological-technological systems with a comparative approach in Latin American and U.S. cities. Cook holds a Ph.D. from Arizona State University and a Bachelor of Arts from Wellesley College.

David M. Iwaniec is an Assistant Professor of Urban Sustainability at the Urban Studies Institute, Andrew Young School of Policy Studies at Georgia State University. He is a sustainability scientist researching anticipatory and systems approaches to advance urban sustainability, resilience, and justice. His work focuses on the co-development of scenarios and transition pathways for positive futures of urban transformation.

Tischa A. Muñoz-Erickson is a Research Social Scientist in the USDA Forest Service's International Institute of Tropical Forestry, in Río Piedras, Puerto Rico. She studies urban sustainability governance, including the policy networks, knowledge systems, anticipatory capacities, and strategies to advance sustainability, resilience, and equity. She is also actively involved in transdisciplinary platforms to facilitate the co-production of futures and transition pathways in the US and Latin American cities. Muñoz-Erickson leads the International Urban Field Station and the San Juan ULTRA and is also Co-PI of the National Science Foundation Urban Resilience to Extreme Events Sustainability Research Network (UREx SRN) and the SETS Convergence project.

Vivian Hobbins is an interdisciplinary scientist working as an Adjunct Faculty in the College of Science for Grand Canyon University. She graduated with honors from Sonora Technological Institute (ITSON) with a BS in Civil Engineering. She also completed an MS in Natural Resources and her Ph.D. in Ecohydrology at ITSON. Her collaborative research with the Urban Resilience to Extremes Sustainability Research Network focused mainly on data acquisition and modeling of the effects of flooding and drought in urban centers in the Americas. Vivian has experience collaborating with various stakeholders in US and Latin American cities to assess the vulnerability of communities and infrastructure to extreme climate events and to develop solutions to build climate resilience. She is now dedicating her life to teaching the next generation of environmental scientists and engineers, as well as empowering instructors to transfer their knowledge and expertise effectively across generations.

Darin Wahl is a Ph.D. candidate at the Lund University Centre for Sustainability Studies. His current work focuses on sustainability innovations in natural resource systems through Urban Living Labs. His research interests include urban resilience to climate change impacts and sense of place in the context of sustainability transitions and transformations. Darin works through interdisciplinary and transdisciplinary processes aiming to learn with others to develop solutions to complex sustainability issues.

Chapter 9
Modeling Urban Futures: Data-Driven Scenarios of Climate Change and Vulnerability in Cities

L. Ortiz, A. Mustafa, B. Rosenzweig, Rocio Carrero, and Timon McPhearson

Abstract Cities are complex systems where social, ecological, and technological processes are deeply coupled. This coupling complicates urban planning and land use development, as changing one facet of the urban fabric will likely impact the others. As cities grapple with climate change, there is a growing need to envision urban futures that not only address more frequent and intense severe weather events but also improve day-to-day livability. Here we examine climate risks as functions of the local land use with numerical models. These models leverage a wide array of data sources, from satellite imagery to tax assessments and land cover. We then present a machine-learning cellular automata approach to combine historical land use change with local coproduced urban future scenarios. The cellular automata model uses historical and ancillary data like existing road systems and natural features to develop a set of probabilistic land use change rules, which are then modified according to stakeholder priorities. The resulting land use scenarios are evaluated against historical flood hazards, showcasing how they perform against stakeholder expectations. Our work shows that coproduced scenarios, when grounded with historical and emerging data, can provide paths that increase resilience to weather hazards as well as enhancing ecosystem services provided to citizens.

The original version of this chapter was revised by adding Dr. Rocio Carrero to the author group. The correction to this chapter is available at: https://doi.org/10.1007/978-3-030-63131-4_13

L. Ortiz (✉) · A. Mustafa · R. Carrero · T. McPhearson
Urban Systems Lab, The New School, New York, NY, USA
e-mail: ortizl@newschool.edu

R. Carrero
e-mail: carrero.r@gmail.com

B. Rosenzweig
Environmental Science, Natural Sciences and Mathematics, Sarah Lawrence College, Bronxville, NY, USA

T. McPhearson
Cary Institute of Ecosystem Studies, Millbrook, NY, USA

Stockholm Resilience Center, Stockholm University, Stockholm, Sweden

Z. A. Hamstead et al. (eds.), *Resilient Urban Futures*, The Urban Book Series,
https://doi.org/10.1007/978-3-030-63131-4_9

Keywords Urban development scenarios · Land use/cover change modeling · Hazard · Mapping · Climate vulnerability

9.1 Data-Driven Models of Urban Land Use and Climate Hazards

Often, urban development scenarios focus on global climate change impacts on existing infrastructure (Ortiz et al. 2019). Although some studies have explored the impacts of projected urban expansion, traditional approaches are often based on planning documents (Krayenhoff et al. 2018) or negative frameworks of the future (Coumou and Robinson 2013). As cities become more aware of the climate-related challenges ahead, spatially explicit models of heat risk, flooding, and other hazards are developed to address present and expected challenges. This chapter explores applications of data-driven approaches to estimate present and future risks related to weather hazards. The modeling efforts detailed in upcoming sections focus on land cover as a driver of heat and local flood hazards. By leveraging a wide array of data sources, from satellite imagery to tax assessments and land cover, models of present and future risk can be developed based on statistical and physical relationships between the land surface and climatological processes. Finally, land use/cover (LUC) models are introduced as a tool to develop future development scenarios. With LUC as a driver of weather hazard frequency and intensity, the combination of these models can provide stakeholders a tool to not just explore envisioned landscapes but also show their impacts on these risks across spatial and temporal domains.

The sections that follow introduce a modeling framework, detailed through a series of case studies focusing on two sites: San Juan, Puerto Rico and New York, New York. As coastal cities, San Juan and New York are exposed to similar hazards, such as flooding and sea-level rise. Each case study city, however, features vastly different infrastructure and socioeconomic conditions, highlighting the versatility of modeling approaches.

9.2 Land Surface Temperature Projections in Cities

Extreme heat is one of the most hazardous weather events, impacting human health, energy use, and ecosystems. Moreover, global climate models project that climate conditions associated with severe heat will become more intense, more frequent, and longer lasting (Meehl and Tebaldi 2004). By concentrating large populations and infrastructure in relatively small geographical areas, cities are particularly vulnerable to extreme heat. This vulnerability is exacerbated by warmer temperatures observed in cities, a phenomenon known as the urban heat island (UHI).

UHIs are a function of urban modification of the land surface, in turn altering its energy balance (Oke 1982). These modifications include reductions in natural cooling mechanisms like evapotranspiration and radiative cooling, along with the addition of heat sources like air conditioning and traffic. UHIs have also been shown to intensify during periods of extreme heat due to synergies between the surface and atmosphere (Li and Bou-Zeid 2013).

Traditionally, future climate is projected at continental to global scales using General Circulation Models (GCMs). GCMs are physical mathematical models that solve the equations of fluid motion and thermodynamics over the entire planet. However, computing resources limit the spatial resolution of GCMs to the order 10^2 km (km). Their coarse spatial resolution makes the use of GCMs problematic as a source of information on future climate hazards in cities, where physical features exist at less than 1 km scales. In addition, physical processes that occur at finer spatial scales than GCM grid resolution are often heavily parameterized (i.e., estimated using bulk or empirical relationships) or not present at all. Many of these processes are particularly important in urban settings, such as the effect buildings have on temperature and winds, as well as infrastructure (e.g., sewage and slopes) on flood extents. Two broad sets of techniques have been developed to address these shortcomings: dynamical and statistical downscaling.

Dynamical downscaling involves models that solve a similar set of equations as GCMs over a limited area, using GCM output data itself as initial and boundary conditions. In addition to employing higher resolution grids, dynamical downscaling can often represent smaller scale processes explicitly, such as convection, land surface dynamics, and clouds. However, availability of computing resources has traditionally limited this approach to simulations in the order of 1–10 km (i.e., neighborhood to city scales) for a limited number of regions (Kong et al. 2019).

On the other hand, statistical downscaling methods map observations to coarse GCM data to increase spatial resolution by use of additional data (e.g., land cover and weather station data). The main benefit of this approach lies in its low computational cost, with spatial resolution being limited by data availability of observations. With modern satellite imagery, observations often exist at the 0.01 to 0.1 km scale, where signals related to individual buildings can be explicitly resolved.

9.2.1 Surface Temperature Projections at City Scales: New York City Case Study

New York City (NYC), the most populous in the USA, faces challenges related to extreme summer heat. Studies have shown that the geographic distribution of temperatures is not uniform throughout the city (Ortiz et al. 2018). Assessment of heat risks is further complicated by the spatial variability of vulnerable groups in NYC (Rosenthal et al. 2014; Madrigano et al. 2015). This spatial variability necessitates the use of models of heat at fine spatial scales, accounting for the geography of the factors driving both temperature and vulnerability.

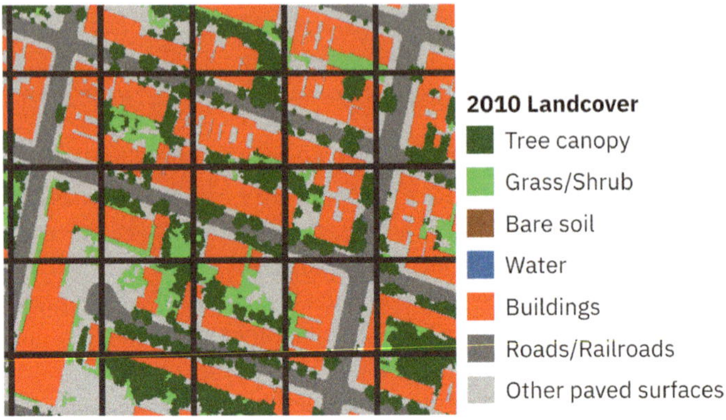

Fig. 9.1 60-m grid used to generate STURLA classes from land cover and building data in New York City

A statistical downscaling approach to derive urban surface temperature futures is to map satellite-derived surface temperatures to urban landscapes. Hamstead et al. (2016) show that by combining land cover with detailed building morphology data, urban landscapes could be classified into statistically separable composite classes with distinct surface temperature distributions. These Structure of Urban Landscape (STURLA) classes are derived by sampling land cover data within adjacent square grids, with each composite class made up of the contained land categories. (For more information about STURLA, see Chap. 4 on vulnerability mapping.) Fig. 9.1 shows a 60 m (m) by 60 m grid over 1 m resolution land cover data in New York City using the 2010 LiDAR-derived land cover dataset. Present-day land surface temperature (LST) is estimated from Landsat 8 thermal imagery as described in Avdan and Jovanovska (2016).

The median LST for each class is then assigned to all cells of that class, so that there is a single LST value mapped to each. Once surface temperatures are mapped to present day imagery, a GCM ensemble is used to project future values. Rather than rely on a single global model as the source of future projections, a common approach is to use a group of simulations from different models. This approach addresses the uncertainty in the assumptions used in GCM formulations and their initial and boundary conditions. Here, an eight-member ensemble is used from the fifth Climate Model Intercomparison Project (CMIP5, Taylor et al. 2012) for an end of century (2070 to 2100) high emissions scenario (Representative concentration pathway 8.5, or RCP 8.5, Riahi et al. 2011). To project future LST, the composite land class-mapped LST is scaled using statistical standardization:

$$LST_{standard} = \frac{LST_i - \overline{LST}}{\sigma_{LST}}$$

where LST_icorresponds to the composite class-mapped temperature values and LST is the mean value, and σ_{LST} *is the LST* standard deviation. LST is then rescaled to using the GCM ensemble's 25th, 50th, and 75th quantiles to show the range of projections available:

$$\Delta LST = \Delta LST_{GCM} + \left(\frac{\Delta LST_{GCM}}{\sigma_{GCM}}\right) * LST_{standard}$$

Results (Fig. 9.2) provide insights into how the various composite urban landscapes may drive temperature change locally as global temperatures increase. For example, LST in relatively flat locations with little vegetation, such as airports, may warm more than in tree-lined areas with few to no buildings (e.g., Central Park). These fine-scale projections can be used to map future heat-related risk and vulnerability, as they provide information at spatial scales close to the size of housing units.

One limitation of this method is that physical processes at the land surface are not explicitly modeled, and thus nonlinear interactions between them may be underestimated or not accounted for at all. For example, droughts and dry soil conditions have been shown to greatly increase the intensity and frequency of heat waves (Fischer et al. 2007). Another limitation is that no temporal land use dynamics are considered. However, as LST change is mapped to specific classes, this method can be coupled with land cover models to explore their impacts on LST. With that approach, specific adaptations to increasing temperatures may be explored, offering stakeholders a measure of how adaptation measures may modify surface conditions and thus impact exposure to heat. These kinds of methods enable analysis of urban

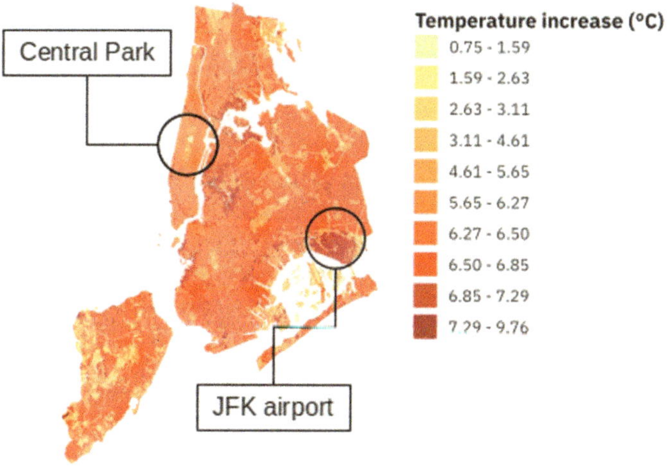

Fig. 9.2 End of century (2070–2100) landscape-mapped surface temperature projections based on the eight-member General Circulation Model RCP8.5 emissions scenario ensemble median values

heat. By tying landscapes to surface temperatures, impacts of urban transformations may be explored, either in the form of heat adaptation or urban development.

9.3 Urban Flooding

Flooding poses an increasing threat worldwide, causing disruption, economic damages and loss of life (Jha et al. 2012). Contemporary cities face risk from multiple types and combinations of meteorologically driven flooding (Moftakhari et al. 2017). These include types that have been well studied, such as fluvial and coastal flooding, along with flooding types that have only recently begun to receive attention, such as pluvial and groundwater flooding (Box 9.1).

Box 9.1 Types of flooding events and their definitions

Flooding types

Fluvial/Lakeshore Flooding: Flooding that occurs when the stage of rivers, streams or other freshwater bodies rises above bankfull elevations and/or the height of levees or flood protection infrastructure.

Coastal Flooding: Flooding that occurs when tide levels exceed an elevation threshold that results in the inundation of infrastructure or disruption of socioeconomic activities.

Pluvial Flooding: Flooding that occurs when precipitation rates exceed the rate of natural or engineered drainage, resulting in overland inundation and/or flow.

Groundwater Flooding: Flooding that occurs when groundwater tables rise above a threshold level that results in the inundation of infrastructure or disruption of socioeconomic activities.

Due to global climate change, the frequency and intensity of all of these flooding mechanisms are projected to increase in many regions of the world (Rotzoll and Fletcher 2013; Arnell and Gosling 2016; Vitousek et al. 2017; Rosenzweig et al. 2018). However, climatic drivers will interplay dynamically with land use pathways to dynamically determine flood risk. LUC models can be used to enhance understanding of how land use changes can impact all three components of flood risk, outlined below (Crichton 1999; Koks et al. 2015).

- Exposure: The population, property and assets located in inundated areas.
- Social vulnerability: The sensitivity of social, ecological and infrastructure systems to the impacts of inundation when and where it occurs
- Hazard: The probability that inundation will occur.

Many studies have used LUC models to assess flooding exposure. This is typically done using a decoupled approach, where future land use and inundated area are simulated independently and the results then compared (Barredo and Engelen 2010; Cammerer et al. 2013; Beckers et al. 2013; Song et al. 2017). While these decoupled approaches can provide valuable information to support urban planning, it is important to consider that cities are integrated social–ecological–technological systems and that “social” land-use planning decisions will determine not just who is living in potentially inundated areas but will also affect the area inundated in response to a meteorological or climatological hazard. For example, the dense encroachment of buildings onto the floodplain can affect exposure during a flooding event (Fig. 9.3).

LUC models can also be used to enhance understanding of social vulnerability to flooding under different scenarios. For any given population or settlement exposed to flooding, the severity of impact is associated with social parameters such as wealth, type of infrastructure, and/or the availability of insurance or other social instruments that support recovery from inundation. For example, informal settlements are known to be highly vulnerable to flooding in the absence of specialized mitigation efforts (Jiusto and Kenney 2016). LUC models can be used to assess the effectiveness of policies to disincentivize the expansion of informal settlements in the floodplain, thus reducing future social vulnerability to flooding (Inouye et al. 2015).

Fig. 9.3 New York City’s East Harlem/Randalls Island community with the event of 1.9 m (75 inches) of sea-level rise. In this scenario, 11% of this community would be inundated during a typical daily high tide (Mean Highest High Water, or MHHW). Through the Urban Resilience to Extremes Sustainability Research Network (URExSRN), researchers are using cellular automata to investigate exposure under alternative land use scenarios

Finally, several studies have identified the potential impact of urban land use development on local meteorological hazards, those of which may result in urban flooding. Conventionally, meteorological hazards were seen as external, independent drivers of flooding. However, recent studies have found that urban canopy and heat island (Lin et al. 2011; Han et al. 2014) effects can impact the evolution and rainfall of both thunderstorms and longer duration extreme rain events, such as Hurricane Harvey in 2018 (Zhang et al. 2018). LUC models can be used to better understand these feedbacks and the integrated dynamics of land use and climate change that can determine future flood risk.

9.4 Modeling Future Land Use/Cover Change Scenarios

Recently, envisioning positive futures has gained traction as a method of designing desirable outcomes in cities (see Chap. 6; McPhearson et al. 2016). This visioning process gives stakeholders, policymakers, and communities an opportunity to set goals and transitions that not only make urban growth more sustainable and resilient to climate hazards but also improve the services they provide. One way to make the visioning processes spatially explicit is through spatially explicit mathematical land use models.

LUC models seek to understand the drivers of LUC dynamics (Mustafa et al. 2018b) and/or simulate possible future scenarios (Hyandye and Martz 2017). Several modeling approaches have been proposed to analyze land change patterns. Broadly, these approaches are cellular automata (CA), agent-based (AB), and statistical models. Among these, CA has been widely used due to its simplicity, explanatory power, and ability to represent LUC evolution (Troisi 2015). The CA framework (Basse et al. 2014; Hyandye and Martz 2017; Mustafa et al. 2018a) is a spatially explicit model in which the change from one land use to another (e.g., from forest to urban) is controlled by the states of neighboring locations (called *cells* in this context). Although pure CA models cannot account for important global LUC change drivers (e.g., distance to roads and slope angles), newer approaches have coupled them to statistical models (e.g., logistic regression) in order to include their influence. AB models (Zhuge et al. 2016; Mustafa et al. 2017) allow the exploration of interactions between different spatial scales (e.g., urban developers and the environment, Mialhe et al. 2012). These models incorporate individuals' behavior and their interactions in the land change process.

Subsequent subsections detail a case study of the use of CA models as a tool to coproduce spatially explicit visions of the future for the Caribbean city of San Juan, Puerto Rico. In this study, we employ the Dinamica Environment for Geoprocessing Objects (EGO), a CA-based model, to simulate possible future land use scenarios of San Juan, Puerto Rico. Unlike typical CA models that use descriptive logistic regression or other static methods to calibrate the relationship between land use change process and its drivers, Dinamica uses the weight of evidence (WoE) method, which has been shown to offer more flexibility in modeling these relationships (Kolb et al. 2013; Pathirana et al. 2014; Gago-Silva et al. 2017). WoE methods use the concept of conditional probability to estimate the weight given to all driving variables as they occur (or not) in historical datasets. This has the effect of modifying the direct impact of each dataset on LUC change, with this weighting being updated with new data. As reproducing these relationships is crucial to simulate LUC change dynamics, Dinamica has been widely employed in this domain.

9.4.1 Land Use/Cover Scenarios Modeling: San Juan, Puerto Rico Case Study

Following the coproduction framework detailed in Chap. 6, we developed three distinct, long-term future (2080) visions of the coastal city of San Juan, Puerto Rico: Food & Energy Security, Coastal and Flooding, and Connected Cities. Each scenario's objectives and priorities were used to modify the conversion rates of respective land cover types using a CA model trained on historical data, as detailed in Table 9.1. These objectives and goals were developed via a series of activities, which included participatory mapping and development of timelines and milestones for each scenario.

In the San Juan case study, the CA model is trained on two LUC datasets: 1991 and 2000. The LUC data, at 10 m spatial resolution, have been reclassified

Table 9.1 Coproduced future scenarios for San Juan, Puerto Rico with their objectives and corresponding cellular automata model rule modifications

Scenario name	Objectives	Modeling transition rules
Food and energy security	Ecotone restoration (wetland and riverine) Use of vacant land for urban and periurban agriculture	Increase conversion to wetlands near coast by 10% Increase conversion to forest near water land use cells Generate small agriculture patches within urban areas
Flooding	Reforestation Relocation of coastal communities to inland locations to strengthen coastal ecosystems and reduce flood risk, starting from 2050	Increase conversion to forest throughout entire domain From 2050 onward, decrease urban areas near coast Increase conversion rates from bare soil to forests
Connected cities	Increase connectivity of transportation infrastructure Rivers and lakes as part of transportation system Reforestation near coast and other water bodies to restore watershed Reduced development near coast	Increase conversion to forest throughout entire domain From 2050 onward, decrease development of urban areas near coast Increase conversion to forest near inland water bodies

into 10 categories: Sea, High-density urban, Low-density urban, Cultivated lands, Pasture, Forests, Wetlands, Coastal sand, Bare soil, and Inland water. In addition, several global drivers of LUC change are included in the model: distances to barrios, road network, airport, vial, lakes, ports, and rivers, as well as protected zones and floodplains.

The change rate from one LUC to another per time step, representing 1 year, is obtained in the CA model by a cross-tabulation between the two LUC maps. Transition rules used to allocate LUC change consists of two components. The first is calculated using the LUC change global drivers. The second component is based on the local neighbors of each cell. Dinamica calculates the transition probability based on global drivers using the WoE method.

After calculating the transition probabilities based on the explanatory variables, Dinamica uses CA model to calculate transition probabilities according to the immediate neighbors for each cell. This is done using two complementary functions: Expander and Patcher. Along with mimicking local neighborhood influence, these functions allow for controlling the geometry of the simulated patches by estimating the mean size, size variance, and isometry of the patches.

9.4.2 San Juan Simulation Results

Simulation results reveal significant differences between the scenarios (Fig. 9.4), consistent with their corresponding stakeholder-stated objectives. In addition to the three coproduced future scenarios, a "business-as-usual" (BAU) scenario was also generated. Development of BAU followed the same modeling approach detailed above, but without any modification of land transition rules, representing a projection of future San Juan based entirely on historical LUC change.

In the Food and Energy Security scenario, green corridors appear along rivers (forest and cultivated patches), with wetland increasing near riverbeds and coastal areas by 2080. In addition, urban development is characterized by a low rather than high density urban fabric, which is predominant in the BAU scenario.

The flooding scenario shows massive reforestation and a relocation of coastal communities. This relocation is coupled with development of catchments to reduce flooding vulnerability, one of the stated scenario goals. This reduced flooding exposure is evident when overlaying the modeled LUC scenarios with the FEMA 500-year floodplain. Total urban area exposed to flooding by the year 2080 is lowest in the Flooding scenario.

The 2080 Connected City simulation is mainly characterized by a pattern of urbanization (including high-density urban) integrated with an increase in green space. The outcome is largely urbanized, but with many corridors and patches of green cover, wetlands and riverine forest (Fig. 9.5).

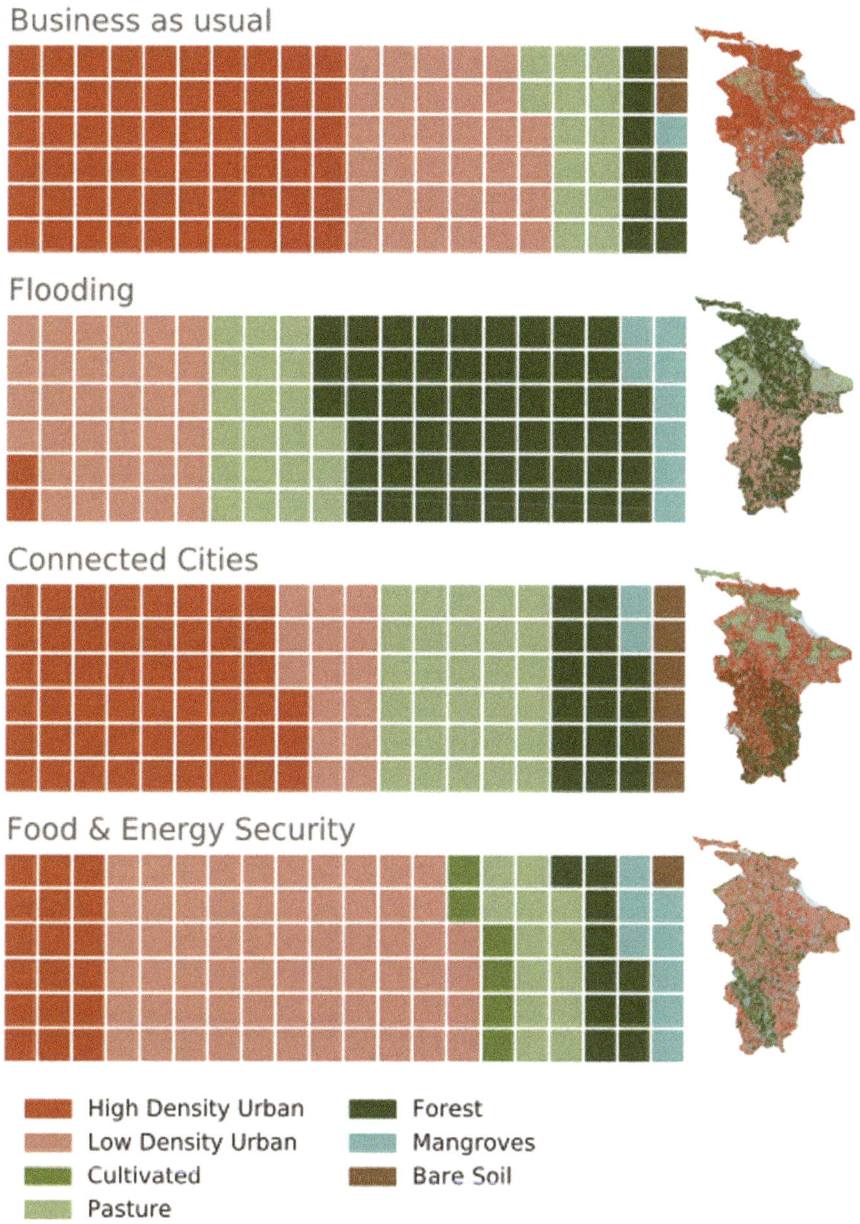

Fig. 9.4 Modeled land use/cover change scenarios. The gridded waffle diagram shows the relative composition by land use/cover category

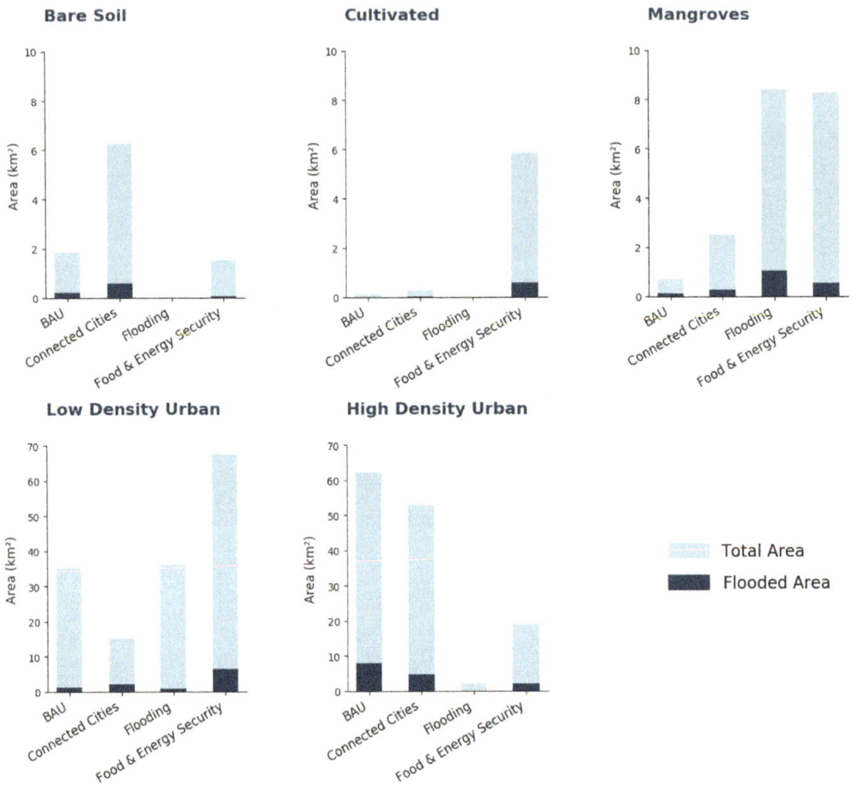

Fig. 9.5 Total and flooded area for selected land use/cover categories in the business as usual and three coproduced land use/cover scenarios

9.5 Conclusion

Weather hazards—projected to become more frequent and intense—pose a large threat to cities and the people in them. The complexity and scale of these threats will require adaptation strategies of commensurate scale. As cities grapple with these challenges, land development will need to account for not just how land use affects services, but how it may also drive future hazards. In this chapter, we showed data-driven approaches to estimate the impacts of LUC on heat and flood risks. These approaches, coupled with the use of CA models, can provide urban planners and policymakers with the tools to not just develop LUC plans, but to also fully explore their impacts on heat and flood risks.

Coupled with participatory production of future visions, these models can provide tools needed to explicitly identify goal tradeoffs. For example, relocating coastal communities may protect them from flood risk, but may also expose them to higher temperatures. Explicitly modeling LUC change can also form an iterative co-production process, where potential scenarios and associated impacts are modified to minimize negative impacts while optimizing positive outcomes.

As data collection efforts increase, so will the utility of these models. Future models may include risks to urban infrastructure such as exposure of electric substations to flood plains in present and future scenarios, or extreme heat in underground subway stations. More elaborate models might include risks of combined hazards (e.g., heat waves followed by power failure due to a flood event), and allow stakeholders to build resilience to them into planning efforts.

With global climate change increasing the severity and frequency of many types of weather hazards, tools to study risks related to the spatial features of cities will be of increasing necessity. Spatially explicit mapping of these hazards will allow policy to address impacts across populations and account for vulnerability in new ways.

References

Arnell NW, Gosling SN (2016) The impacts of climate change on river flood risk at the global scale. Clim Change 134(3):387–401. https://doi.org/10.1007/s10584-014-1084-5

Avdan U, Jovanovska G (2016) Algorithm for automated mapping of land surface temperature using LANDSAT 8 satellite data. J Sensors 2016:1–8. https://doi.org/10.1155/2016/1480307

BarredoJI EG (2010) Land use scenario modeling for flood risk mitigation. Sustainability 2(5):1327–1344. https://doi.org/10.3390/su2051327

Basse RM, Omrani H, Charif O et al (2014) Land use changes modelling using advanced methods: cellular automata and artificial neural networks. The spatial and explicit representation of land cover dynamics at the cross-border region scale. ApplGeogr 53:160–171. https://doi.org/10.1016/j.apgeog.2014.06.016

Beckers A, Dewals B, Erpicum S et al (2013) Contribution of land use changes to future flood damage along the river Meuse in the Walloon region. Nat Hazards Earth Syst Sci 13:2301–2318. https://doi.org/10.5194/nhess-13-2301-2013

Caldwell P, Chin H-NS, Bader DC et al (2009) Evaluation of a WRF dynamical downscaling simulation over California. Clim Change 95(3–4):499–521. https://doi.org/10.1007/s10584-009-9583-5

Cammerer H, Thieken AH, Verburg PH (2013) Spatio-temporal dynamics in the flood exposure due to land use changes in the Alpine Lech Valley in Tyrol (Austria). Nat Hazards 68(3):1243–1270. https://doi.org/10.1007/s11069-012-0280-8

Coumou D, Robinson A (2013) Historic and future increase in the global land area affected by monthly heat extremes. Environ Rese Lett 8(3):034018. https://doi.org/10.1088/1748-9326/8/3/034018

Crichton D (1999) The Risk Triangle. Nat Disaster Manage 102:103

Fischer EM, Seneviratne SI, Vidale PL et al (2007) Soil moisture–atmosphere interactions during the 2003 European summer heat wave. J Clim 20(20):5081–5099. https://doi.org/10.1175/JCLI4288.1

Gago-Silva A, Ray N, Lehmann A (2017) Spatial dynamic modelling of future scenarios of land use change in Vaud and Valais. Western Switzerland. ISPRS Int J Geo-Inf 6(4):115. https://doi.org/10.3390/ijgi6040115

HamsteadZA KP, Larondelle N et al (2016) Classification of the heterogeneous structure of urban landscapes (STURLA) as an indicator of landscape function applied to surface temperature in New York City. Ecol Indic 70:574–585. https://doi.org/10.3390/su10030645

Han J-Y, Baik J-J, Lee H (2014) Urban impacts on precipitation. Asia-Pacific J Atmos Sci 50(1):17–30. https://doi.org/10.1007/s13143-014-0016-7
Hyandye C, Martz LW (2017) A Markovian and cellular automata land-use change predictive model of the Usangu Catchment. Int J Remote Sens 38(1):64–81. https://doi.org/10.1080/01431161.2016.1259675
Inouye CEN, de Sousa WC, de FreitasDM, et al (2015) Modelling the spatial dynamics of urban growth and land use changes in the north coast of São Paulo, Brazil. Ocean Coastal Manage 108:147–157. https://doi.org/10.1016/j.ocecoaman.2014.12.016
Jha AK, Bloch R, Lamond J (2012) Cities and flooding: a guide to integrated urban flood risk management for the 21st Century. the World Bank. https://doi.org/10.1596/978-0-8213-8866-2
Jiusto S, Kenney M (2016) Hard rain gonna fall: strategies for sustainable urban drainage in informal settlements. Urban Water J 13(3):253–269. https://doi.org/10.1080/1573062X.2014.991329
Koks EE, Jongman B, Husby TG et al (2015) Combining hazard, exposure and social vulnerability to provide lessons for flood risk management. Environ Sci Policy 47:42–52. https://doi.org/10.1016/j.envsci.2014.10.013
Kolb M, Mas J-F, Galicia L (2013) Evaluating drivers of land-use change and transition potential models in a complex landscape in Southern Mexico. Int J GeogInf Sci 27(9):1804–1827. https://doi.org/10.1080/13658816.2013.770517
Kong X, Wang A, Bi X et al (2019) Assessment of temperature extremes in China using RegCM4 and WRF. Adv Atmos Sci 36(4):363–377. https://doi.org/10.1007/s00376-018-8144-0
Krayenhoff ES, Moustaoui M, Broadbent AM et al (2018) Diurnal interaction between urban expansion, climate change and adaptation in US cities. Nat Clim Change 8(12):1097. https://doi.org/10.1038/s41558-018-0320-9
Li D, Bou-Zeid E (2013) Synergistic interactions between urban heat islands and heat waves: the impact in cities is larger than the sum of its parts. J ApplMeteorolClimatol 52(9):2051–2064. https://doi.org/10.1175/JAMC-D-13-02.1
Lin CY, Chen WC, Chang PL et al (2011) Impact of the urban heat island effect on precipitation over a complex geographic environment in Northern Taiwan. J ApplMeteorolClimatol 50(2):339–353. https://doi.org/10.1175/2010JAMC2504.1
Madrigano J, Ito K, Johnson S et al (2015) A case-only study of vulnerability to heat wave–related mortality in New York City (2000–2011). Environ Health Perspect 123(7):672–678. https://doi.org/10.1289/ehp.1408178
McPhearson T, IwaniecDM BX (2016) Positive visions for guiding urban transformations toward sustainable futures. CurrOpin Environ Sustain 22:33–40. https://doi.org/10.1016/j.cosust.2017.04.004
Meehl GA, Tebaldi C (2004) More intense, more frequent, and longer lasting heat waves in the 21st century. Sci 305(5686):994–997. https://doi.org/10.1126/science.1098704
Mialhe F, Becu N, Gunnell Y (2012) An agent-based model for analyzing land use dynamics in response to farmer behaviour and environmental change in the Pampanga delta (Philippines). AgricEcosyst Environ 161:55–69. https://doi.org/10.1016/j.agee.2012.07.016
Moftakhari HR, AghaKouchak A, Sanders BF et al (2017) Cumulative hazard: the case of nuisance flooding. Earth's Future 5(2):214–223. https://doi.org/10.1002/2016EF000494
Mustafa A, Cools M, Saadi I et al (2017) Coupling agent-based, cellular automata and logistic regression into a hybrid urban expansion model (HUEM). Land Use Policy 69:529–540. https://doi.org/10.1016/j.landusepol.2017.10.009
Mustafa A, Heppenstall A, Omrani H et al (2018) Modelling built-up expansion and densification with multinomial logistic regression, cellular automata and genetic algorithm. Comput Environ Urban Syst 67:147–156. https://doi.org/10.1016/j.compenvurbsys.2017.09.009
Mustafa A, Van Rompaey A, Cools M et al (2018) Addressing the determinants of built-up expansion and densification processes at the regional scale. Urban Stud 55(15):3279–3298. https://doi.org/10.1177/0042098017749176
OkeTR, (1982) The energetic basis of the urban heat island . Q J R MeteorologSoc 108(455):1–24. https://doi.org/10.1002/qj.49710845502

Ortiz LE, González JE, Horton R et al (2019) High-resolution projections of extreme heat in New York City. Int J Climatol 39(12):4721–4735. https://doi.org/10.1002/joc.6102
Ortiz LE, Gonzalez JE, Wu W et al (2018) New York city impacts on a regional heat wave. J ApplMeteorolClimatol 57(4):837–851. https://doi.org/10.1175/JAMC-D-17-0125.1
Pathirana A, DenekewHB VW et al (2014) Impact of urban growth-driven landuse change on microclimate and extreme precipitation—a sensitivity study. Atmos Res 138:59–72. https://doi.org/10.1016/j.atmosres.2013.10.005
Riahi K, Rao S, Krey V et al (2011) RCP 8.5—a scenario of comparatively high greenhouse gas emissions. Clim Change 109:33. https://doi.org/https://doi.org/10.1007/s10584-011-0149-y
Rosenthal J, Kinney PL, Metzger KB (2014) Intra-urban vulnerability to heat-related mortality in New York City, 1997–2006. Health Place 30:45–60. https://doi.org/10.1016/j.healthplace.2014.07.014
Rosenzweig BR, McPhillips L, Chang H et al (2018) Pluvial flood risk and opportunities for resilience . Wiley Interdiscip Rev: Water 5(6):e1302. https://doi.org/10.1002/wat2.1302
Rotzoll K, Fletcher CH (2013) Assessment of groundwater inundation as a consequence of sea-level rise. Nat Clim Change 3(5):477–481. https://doi.org/10.1038/nclimate1725
Song K, Kwon N, Anderson K et al (2017) Predicting hourly energy consumption in buildings using occupancy-related characteristics of end-user groups. Energy Build 156:121–133. https://doi.org/10.1016/j.enbuild.2017.09.060
Taylor KE, Stouffer RJ, Meehl GA (2012) An overview of CMIP5 and the experiment design. Bull Am MeteorolSoc 93(4):485–498. https://doi.org/10.1175/BAMS-D-11-00094.1
Troisi A (2015) Can CA describe collective effects of polluting agents? Int J Mod Phys C 26(10):1550114. https://doi.org/10.1142/S0129183115501144
Vitousek S, Barnard PL, Fletcher CH et al (2017) Doubling of coastal flooding frequency within decades due to sea-level rise. Sci Rep 7(1):1399. https://doi.org/10.1038/s41598-017-01362-7
Zhang W, Villarini G, Vecchi GA et al (2018) Urbanization exacerbated the rainfall and flooding caused by hurricane Harvey in Houston. Nat 563(7731):384. https://doi.org/10.1038/s41586-018-0676-z
Zhuge C, Shao C, Gao J et al (2016) Agent-based joint model of residential location choice and real estate price for land use and transport model. Comput Environ Urban Syst 57:93–105. https://doi.org/10.1016/j.compenvurbsys.2016.02.001

Luis Ortiz is a postdoctoral research fellow with the Urban Systems Lab at The New School. A mechanical engineer by training, he is interested in the intersection of built environments, humans, and the atmosphere. In particular, he is interested in how climate change impacts energy use and health, as well as how cities may mitigate and adapt to these impacts. He uses both physical and statistical techniques to model feedbacks between buildings and the air around them and how they might change under a warming climate. His work on extreme heat projections was featured in the New York City Panel on Climate Change third report, where he co-authored the chapter on advances in climate science.

Ahmed Mustafa is a spatial data scientist with expertise in urban studies. His research focuses on using methodological approaches that work with complex data sources to study urban systems. He is currently working on developing and implementing computational models to simulate spatial patterns of future cities within the NSF UREx SRN and the NSF Convergence SETS for Urban Resilience projects.

Bernice R. Rosenzweig is an Assistant Professor of Environmental Science at Sarah Lawrence College in Bronxville, New York. Her research investigates the resilience of urban socio-technical systems to extreme meteorological events. Her current work focuses on post-industrial cities in the northeast United States.

Rocio Carrero is a Faculty Fellow at the Urban Systems Lab and an environmental geoscientist interested in testing experimental models to explore urban resilience. As part of Urban Resilience to Extremes (UREx) network's Scenario Modeling Task Force, she worked on developing computational Cellular Automata models to spatially simulate future scenarios.

Timon McPhearson is Director of the Urban Systems Lab and Associate Professor of Urban Ecology at The New School. He is a Research Fellow at The Cary Institute of Ecosystem Studies and Stockholm Resilience Centre and a member of the IPCC.

Chapter 10
Visualizing Urban Social–Ecological–Technological Systems

Daniel Sauter, Jaskirat Randhawa, Claudia Tomateo, and Timon McPhearson

Abstract The Urban Systems Lab (USL) Dataviz Platform is an interactive web application to visualize Social, Ecological, and Technological Systems (SETS). This platform is being used to encourage participatory processes, produce new knowledge, and facilitate collaborative analysis within nine Urban Resilience to Weather-related Extremes (UREx) Sustainability Research Network cities. It allows seamless shifts across contexts, scales, and perspectives for analysis within the SETS framework. How is digital space conceptualized for urban analysis and interventions? What is the capacity for reciprocal relationships between digital and physical space? How do we visually understand urban systems and complex spatial relationships? This chapter provides a comprehensive overview of the application stack and the different representational categories embedded in the Dataviz Platform. Offering a common visual language to various stakeholders, we explore new ground as we believe it has the potential to change how we think about, plan, and design our cities. ("Map devices such as a frame, scale, orientation, projection, indexing and naming reveal artificial geographies that remain unavailable to human eyes." (Corner,.Cosgrove (ed), Mappings, Reaktion Books, London, 1999)

Keywords Spatial narratives · Social-ecological-technological systems (SETS) · Data visualization · Web-based visualization · Resilience

10.1 The USL Dataviz Platform

Facilitated by a UREx team of academic researchers, city planners, first responders, NGOs, and local community groups, the 3D visualizations featured on the

D. Sauter (✉) · J. Randhawa · C. Tomateo · T. McPhearson
The New School, Urban Systems Lab, New York, NY, USA
e-mail: sauter@newschool.edu

T. McPhearson
Cary Institute of Ecosystem Studies, Millbrook, NY, USA

Stockholm Resilience Center, Stockholm University, Stockholm, Sweden

Z. A. Hamstead et al. (eds.), *Resilient Urban Futures*, The Urban Book Series,
https://doi.org/10.1007/978-3-030-63131-4_10

USL Dataviz Platform (Sauter et al. 2019) are used to support cross-city analysis, collaboration, and conversation, with an individualized focus on local priorities in each city. For example, the Dataviz Platform has been used for stakeholder engagement in a number of UREx scenario workshops, providing the spatial context and SETS data to explore desirable and plausible pathways for reducing risks in improving resilience of vulnerable communities and critical infrastructure in cities such as Phoenix, Portland, Miami, and San Juan.

The platform is a public repository for a variety of visual outputs in one integrated project platform. It serves as an engagement tool and supports seamless shifts in context and perspective for analysis within a SETS framework (see Chap. 3); for instance, by switching from geospatial maps focused on heat (Fig. 10.1) to those focused on flood risk, and from city-wide analysis to block-level views. For the sake of inclusivity, common digital and mobile devices can be used to view the intricacies of geospatial layers presented on the platform in a simple and rich way. Additionally, the platform leverages common visual assets from location-based services and maps on mobile devices to relate to user expectations.

UREx city teams have used the platform to draft narratives in a content management system and share them publicly. For example, diverse stakeholders in Syracuse, New York have been able to use the platform to view important SETS data of the city together. This collaboration provides a basis for exploration of resilience futures, as well as a starting point for asking questions about the systemic interactions that drive current challenges in the city. The platform enables teams to manage the publication life cycle and add context and metadata to complex visual outputs such as future heat, flood, and land-use projections (e.g., UREx 2050 and 2080 scenarios). Doing

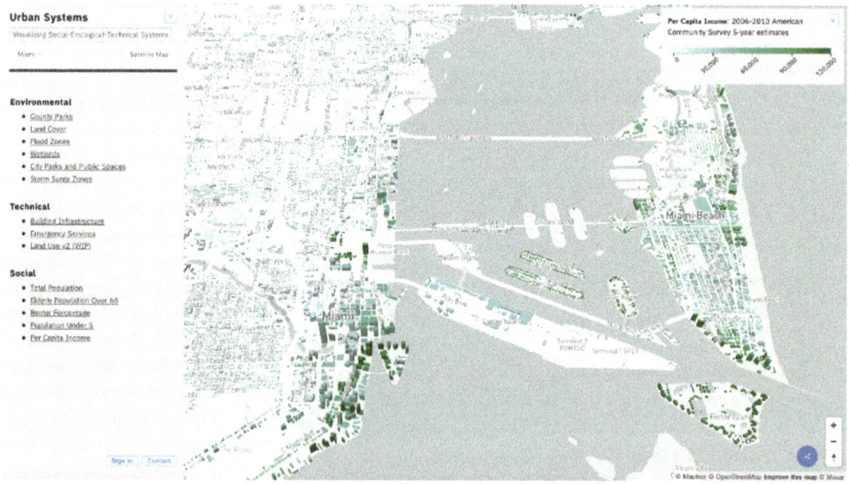

Fig. 10.1 Per capita income in Miami layered onto 3D built infrastructure emphasizes demographic distribution along the coast

so provides a better understanding of the underlying data and computational models that generated a particular output.

10.2 Representation of Space

In *Social Justice and the City* (1973), David Harvey argues the importance of reflecting on the nature of spaces to understand urban processes—what we call today "urban systems." We use Harvey's spatial classification from "Space as a Key Word" (2004) as a vocabulary to name the contextual and perspective shifts featured within the online USL Dataviz Platform:

- Absolute, a fixed space, true to reality, one frame, independent
- Relative, multiple geometries to choose from, "spatial frame depends upon what is being relativized and by whom" (Harvey 2004, p. 3)
- Relational, processes occur and define their own spatial frame, object exists only in the relationship with another object

An interactive platform holds a dynamic range of these spaces (absolute, relative, and relational). Absolute space is represented by building footprints, land use, coastal flooding and everything that can be visualized and measured in space at scale (Fig. 10.2). Relative space is exponentially richer within an interactive tool, because the relativization changes according to user input, zoom levels, and data resolution. Because space is dynamic, data is relative to its container: pixels. While navigating

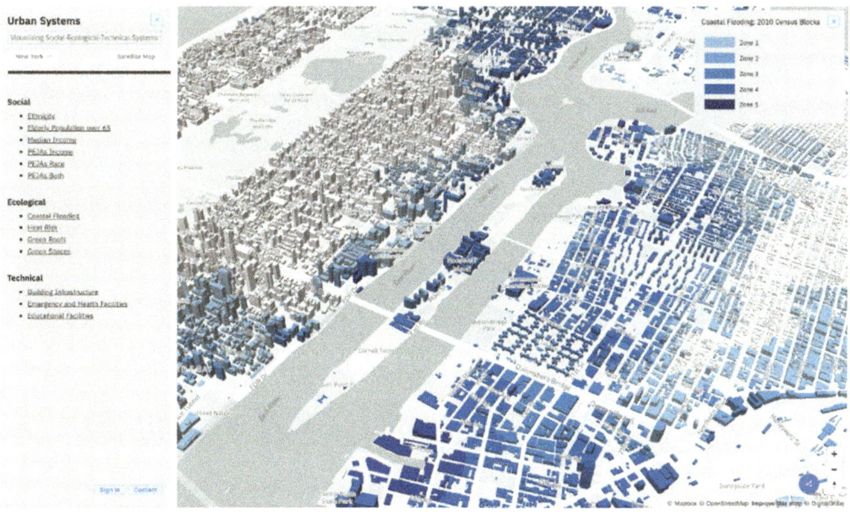

Fig. 10.2 Coastal flooding zones in New York City layered onto 3D built infrastructure emphasizes both social and infrastructural risk associated with potential coastal flooding

the platform, data gets sorted, selected, abstracted, and slowly unpacked to add detail as the user moves closer in.

Data are meaningful on a screen insofar as it is visualized in its relation to other layers at different levels of abstraction. Sometimes the accuracy of elements in an (absolute) geospatial space is secondary; e.g., by using extruded building footprints as an approximation of a building's shape. In other instances, visual elements represent data beyond its (relative) location; e.g., by showing the flow of people, a volume of water, or associated costs. In contrast, relational space is independent from absolute and relative dimensions, operating at the highest level of abstraction based on its own design rules; e.g., diagrams showing stakeholder networks, labor processes, or icons signifying sensations.

A good example for a dataset that shifts from a relative to absolute space paradigm is the green roofs layer in New York City (NYC); zooming out, we can see circles with some transparency showing the green roofs location in NYC (Fig. 10.3). These circles are not at scale but are able to show a distribution of green roofs across the city, relative to each other. As we zoom in, the circles disappear and transition into absolute space, uncovering both the geolocation and footprint of green roofs, as well as their elevation in three-dimensional space (Fig. 10.4).

Harvey further develops the classification where absolute, relative, and relational space intersects to create a matrix of combinations:

- *Material space*, as in what you can experience with your senses
- *Representations of space*, including diagrams, collages, infographics, writing, and geospatial maps. The visualization platform lies mostly in this category because

Fig. 10.3 New York City green roofs, city-wide view, available at http://urex.urbansystemslab.com/share/1t9wrz

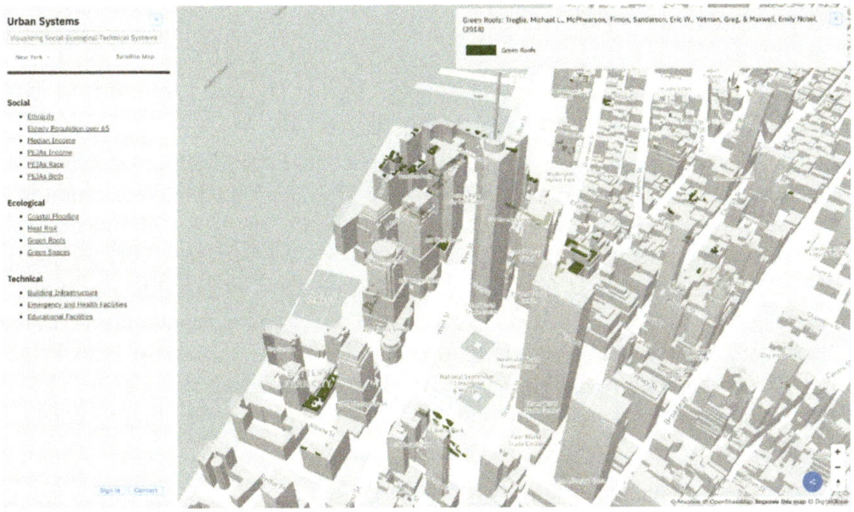

Fig. 10.4 New York City green roofs, neighborhood-level view

it shows data within a specific location on earth, attributing information to the built environment (cities).

- *Spaces of representation*, a symbolic space produced by cultural memories, symbols, emotions, images, and imaginaries such as artistic interpretations and narratives

Considering the different kinds of practices at work in the URExSRN, conceptualization of space varies considerably among the different stakeholders, and spans the full range of Harvey's classification. "The problem of the proper conceptualization of space is resolved through human practice with respect to it.... The question what is space is therefore replaced by the question: How is it that different human practices create and make use of different conceptualizations of space?" (Harvey 1973, pp 13–14).

For example, a hydrologist might look at a particular storm event as a predictable nuisance that occurs at a certain frequency and can be represented accordingly. An emergency responder might view the same event through the emotional account of callers in distress. A resident (materially) experiences the event and its implications from a viewpoint centered at the actual site. And a City official manages those external forces and points of view to develop long-term resiliency plans, using spatial representations such as maps and diagrams to engage and communicate. The different stakeholders and practices therefore require a common (visual) language to bridge those gaps and points of view. This challenge is our objective, organizing a web of spatial relationships and representations into one integrated visualization platform.

10.3 Visualization Concepts

Web-based visualizations operate at the intersection of visual design, statistical analysis, and computer science (Drucker 2014). They create dynamic representations that can shift seamlessly between narrative modes and spatial representations. To balance the two for a range of use cases, the USL Dataviz Platform requires a high level of customization and is therefore unique in how it is constructed from source data to the finished design. Geospatial Information Systems (GIS) are historically used to produce maps for print media, and are tailored towards expert audiences versed in reading geospatial information. The following section discusses the different spatial capacities of the USL Dataviz Platform in relation to spatial categories that are inherent in working with GIS. They are arranged as opposite sides of a representational spectrum, along with a discussion of the design implications for the platform:

- **Static versus Dynamic**: Static GIS maps often function as visual evidence to convey a particular point of view, placed intentionally into a larger narrative by an author. When visualizing data at different scales and levels of fidelity, two factors remain constant: the resolution of the underlying data and the size of the canvas it is displayed within. Hence, there are obvious limitations for constructing complex visualizations where time, space, and scale work interdependently together to produce a visual output within the constraints of the browser canvas. A framework that is flexible enough to shift spatial representations requires an interactive system that can process user inputs and render graphic outputs based on those interactions, dynamically. This approach has the advantageous capacity to do the following: Conduct self-serve analysis on various datasets, display dynamically updated (and real-time) data, facilitate collaboration and dissemination, provide varying levels of access, and collect structured feedback. Additionally, a dynamic system can harness principles of cinematography and motion graphics to tell stories in a linear or nonlinear fashion for different audiences and purposes, such as identifying risks, managing interventions, and planning adaptation strategies (Amini et al. 2015). Such participatory processes are fundamental for an inclusive approach, connecting researchers, practitioners, and residents around particular issues. To that end, a key feature of the platform is the ability to share a specific geospatial view through short URLs, available in the browser (bottom-right) corner, to show the exact same view and settings to a person who opens the link on another device.
- **Global versus Local**: One of the primary affordances of interactive 3D maps is to provide viewers the freedom to seamlessly explore and compare the data at local, regional, national, and global scales. A resident engaged in a local issue, for instance, might therefore better understand how local social-ecological-technological systems, or SETS, compare to a regional picture. At the most local level (fully zoomed in), we use 3D extrusions of building footprints as a visual canvas to display the underlying data. The demographic data we use are accurate at a census block level, and while we extrude each building on the lot, the underlying data are not available at a lot level. Therefore, multiple buildings on the

lot show the same census data. As a representation, building geometry (extruded from footprints using height information) provides a much better sense of the built environment, especially in urban settings, but also suggests higher accuracy—a trade-off and a design decision. Moving from a local to a regional zoom level, the accuracy of building geometry loses significance. Computationally, to display a city-wide view requires the ability to reduce complexity of the data to allow it to be rendered quickly in an online environment. This involves reducing the graphic complexity of buildings shown on the platform as the viewer moves around, thereby balancing the available computational resources and ensuring that the user experience is fluid. For example, showing all buildings on the more than one million building lots of NYC requires significant computation, and in order to find balance, thus requires reducing building detail to maintain a sufficient frame rate and a good user experience across a range of devices, including mobile phones. At a distance, administrative boundaries become more relevant and hold more weight for analysis, and the map geometry adapts accordingly into aggregate bins where necessary.

- **Raster versus Vector**: Vector data format, which is essentially a database of point, line and polygon coordinates, is ideal for representing discrete data and geometries. It plays well with digital applications that demand a suite of interactions on top of the data layer, letting users "touch" and interact with the data itself. On the contrary, working with raster data is suitable for working with continuous data, where working with pixels instead of coordinates is a more computationally sensible approach. Such may be the case for working with digital elevation models (DEM), satellite imagery (Fig. 10.5) and similar remote sensing data. Depending

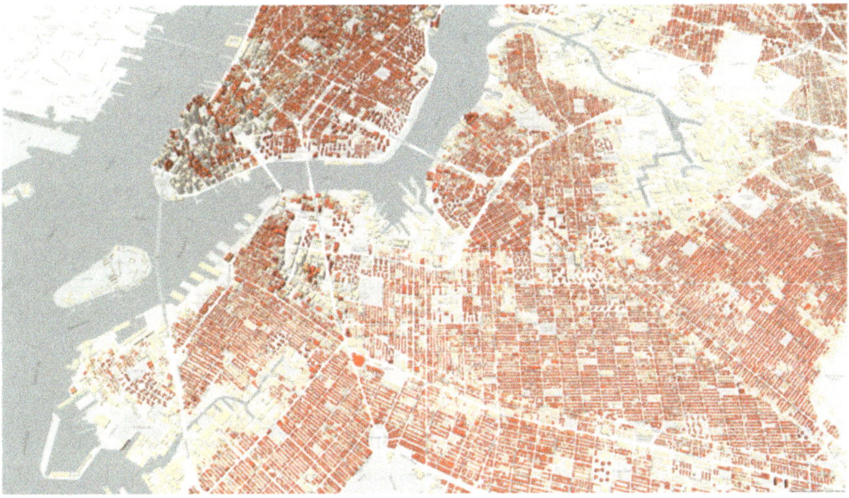

Fig. 10.5 Heat risk in New York City ranging from low to high levels layered onto 3D building infrastructure. (For more information about heat vulnerability mapping, see Chaps. 4 and 9.)

Fig. 10.6 Heat risk in New York City as a spatial overlay on satellite imagery to provide high resolution local context

on the workflow requirements, the data can be converted between raster to vector and vice versa (Fig. 10.6).

- **Narrative versus Representational**: A key decision in the development of the USL Dataviz Platform was the objective to show qualitative data and narrative content developed during stakeholder workshops held in each network city (Fig. 10.7), and to convey connections between various vulnerabilities and exposures; for instance, food and energy security, coastal flooding, and urban and river flooding. Using an online editor, authors can draft narratives in the content management system, embed rich media such as images and movie clips, and link those with geospatial views and layers provided through the platform. Multiple views can be recorded and animated to tell a particular story, presented in a split-screen view that allows for analytic comparisons while navigating the narrative. This design choice balances the qualitative and quantitative representations that are displayed, while keeping the interactive features of the platform active. The design implication is the need for a highly customizable mapping engine,[1] along with the data structures that allow for customizable views, virtual camera settings,

[1] The USL Dataviz Team reviewed a set of mapping libraries to decide on the visualization software stack for the project, including MapBox-GL (which renders buildings as well as building-part extrusions), CesiumJS (using WebGL to render more detailed CityGML geometries), Mapbox SDK for Unity (useful for rendering extruded buildings over a terrain mesh, high-end graphics rendering, and post-processing), along with cloud-based mapping services such as Google Maps, Earth, and ArcGIS Online. Criteria for building a custom platform included the level of customizability, user experience across devices leveraging WebGL, cost to develop, scale, and host a tile server online, available support by a developer community, as well as the graphics roadmap for future development. The decision was to build the platform on Mapbox-GL.

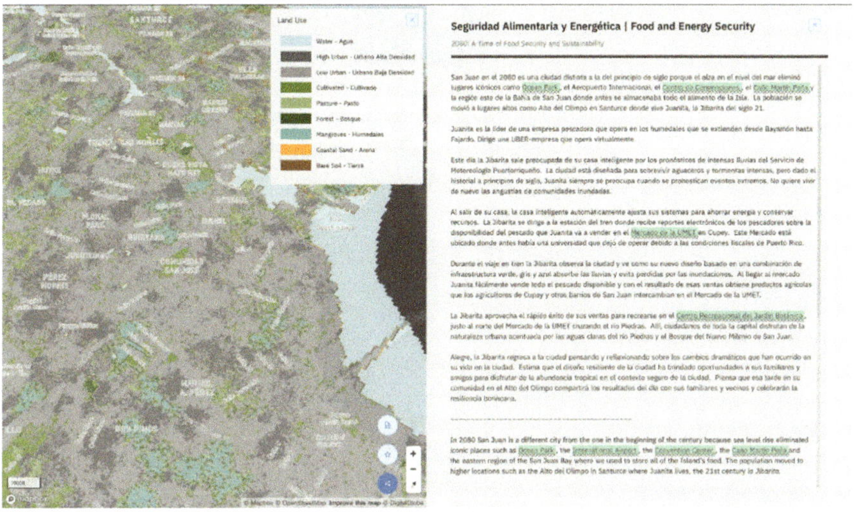

Fig. 10.7 Interactive "Food and Energy Security" qualitative narrative juxtaposed with quantitative geo-spatial-modeled future scenario outputs

user permissions, and metadata. For data representations such as flood risk or land cover, one limitation of the popular Mapbox-GL engine is that it currently does not display elevation as 3D terrain, but as flat images with different color values corresponding to the height of the terrain.

10.4 Application Stack

The USL Dataviz Platform is implemented as a web application using open source libraries, and thus it can be replicated using standard web development tools and best practices.[2] The following workflow diagram broadly illustrates the end to end framework for creating spatial visualizations with the platform, presented as individual layers in the front-end web interface. These layers are interactively transformed client side by viewers through their individual web browsers (local storage) (Fig. 10.8).

The data visualization workflow operates as follows:

- Data preparation: This stage involves collecting the geospatial data and applying necessary transformations, such as adjusting map projection, aggregation, filtering, geocoding, spatial joins, etc. to make the data interoperable with other server-side components for further processing. This is the most meticulous and time-consuming part of the process. The methodology varies highly for each

[2]https://opensource.urbansystemslab.com/.

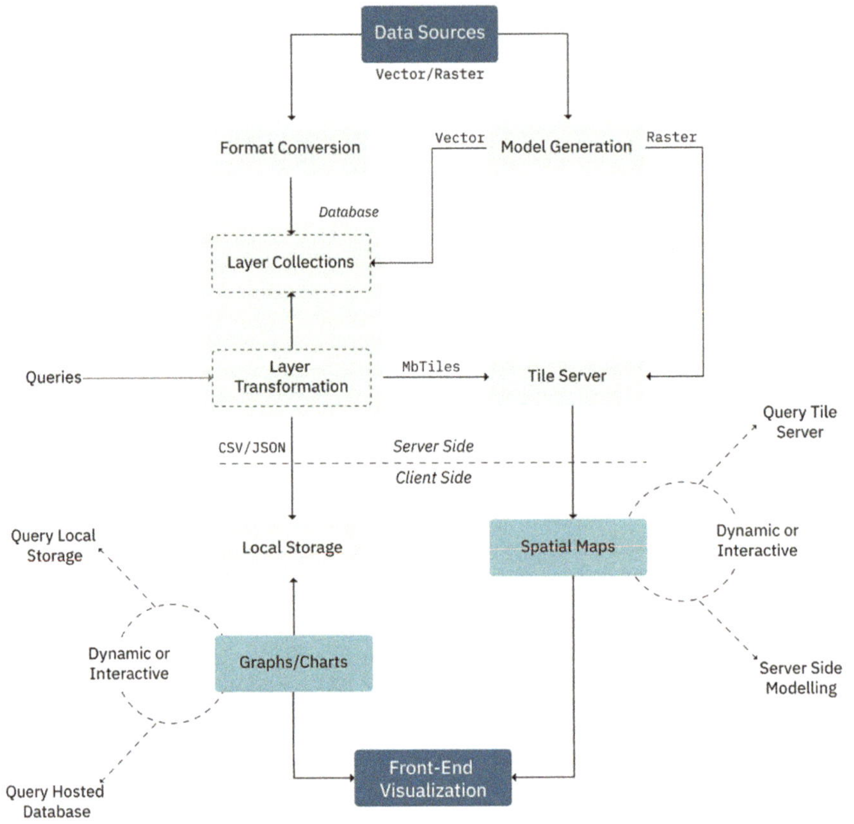

Fig. 10.8 USL Dataviz Platform data preparation and visualization workflow

dataset depending on the programmer's domain expertise and desired output specifications for the visualization, as prioritized by local stakeholders.

- Tile generation: Once the data are converted into the desired raster or vector format, the next step is to generate a tile map from it, generating a set of individual images that are loaded and displayed in the browser as a seamless map. Tippecanoe,[3] a command line utility, is a popular tool that we use to generate tiles from large datasets. It provides extensive options to customize both raster and vector data types.
- Tile server: After encoding the map layers, tiles can be self-hosted on cloud infrastructure or commercial service providers.[4] Many libraries, like D3.js[5]or Deck.gl,[6] require data to be loaded within the client browser. However, streaming

[3]https://github.com/mapbox/tippecanoe.

[4]https://wiki.openstreetmap.org/wiki/Tile_servers.

[5]https://d3js.org/.

[6]https://deck.gl/.

the complete datasets used to a client's device is not feasible. When viewing large datasets in excess of a few megabytes, streaming the data as slippy maps[7] is a standard and preferred approach. The data are split into a grid of tiles for each zoom level. Based on the web browser's viewport, only the tiles within view are requested from the tile server and stitched together into a seamless canvas, which happens while zooming or panning a map. There are a number of tile server solutions available that can be deployed according to usage and demand.[8]
- Front-end application: The data are finally consumed and rendered by a front-end application operating on a user's mobile or desktop device. It is responsible for accepting interactive user inputs and querying the data from the hosted tile server.

10.5 Conclusion

Our objectives are to support the exploration and understanding of complex geospatial relationships in a simple and organized way and to avoid visual outputs that cognitively overwhelm the viewer, both for a variety of use cases. By categorizing the representational concepts that inform our design decisions, the hidden operations within the visualization platform become more transparent. Intentionally, every design decision serves a specific narrative or representational purpose, which might not be apparent at first glance.

As explained by James Corner, maps are artificial geographies that reveal "the invisible" to the human eye. Whether in absolute, relative, or relational space, our interactive platform explores new ground in which maps function as dynamic data aggregates that shift scale and context, reveal transformative narratives, and enable participatory processes. This exploration has the potential to change how we think about, plan, and design our cities.

As a platform designed to provide spatial context and a digital common ground for various stakeholders in group settings, some viewers find the 3D environment more difficult to navigate than others, especially if they are not used to working with spatial data. Along with the spatial shifts from regional to local views, the "artificial geographies" produced by the platform can be cognitively taxing, especially when switching between different SETS data layers. Narrative text, hyperlinks, legends, and sources provide additional information through the platform's narrative panel. This however also requires more in-depth study and sustained engagement in order to process and understand interconnected data.

The impact of the USL data visualization platform lies in its capacity to communicate the invisible through a common visual language, suitable for different human practices and stakeholders. As we continue to research and visualize urban social–ecological–technological systems in the future, we plan on pushing the boundaries

---

[7]https://wiki.openstreetmap.org/wiki/Slippy_Map.

[8]Including Kosmtik, available at https://github.com/kosmtik/kosmtik, and Tileover, available at https://github.com/fl orianf/tileoven.

between absolute and abstract spatial representations in contexts such as social relationships, traffic flows, green infrastructure, and questions of equity.

References

Amini F, Henry Riche N, Lee H et al (2015) Understanding data videos: looking at narrative visualization through the cinematography lens. Proceedings of the 33rd Annual ACM Conference on Human Factors in Computing Systems, Seoul, South Korea, pp 1459–1468

Corner J (1999) The agency of mapping: Speculation, critique and invention. In: Cosgrove D (ed) Mappings. Reaktion Books, London, pp 231–252

Drucker J (2014) Graphesis: visual forms of knowledge production (metaLABprojects). Harvard University Press, Cambridge, MA

Harvey D (1973) Social justice and the city. John Hopkins University Press, Baltimore

Harvey D (2004) Space as a keyword. Presented at the Marx and Philosophy Conference, London: Institute of Education

McPhearson T, Pickett STA, Grimm N et al (2016) Advancing urban ecology toward a science of cities. . Biosci 66(3):198–212. https://doi.org/10.1093/biosci/biw002

Sauter D, Randhawa J, McPhearson T (2019) USL Dataviz Platform. http://nyc.urbansystemslab.com/. Accessed 30 Jun 2020

Daniel Sauter is an Associate Professor of Data Visualization at Parsons School of Design. He serves as the Director of the MS Data Visualization program, Co-Director of the Integrative Ph.D. Fellowship Program, and Associate Director of the Urban Systems Lab at The New School in New York City.

Jaskirat Randhawa is a Lead Experience Manager at Verizon. He is the Product Development Lead at Senscity, a climate intelligence company which he co-founded in 2018. He served as the Technical Lead for data-visualization projects at the Urban Systems Lab from 2016–2019.

Claudia Tomateo is a Research Associate in the Urban Systems Lab at the New School in New York City and Co-Founder of Urban Collective, a design practice for the built environment. As an architect and urban designer, her research focuses on the intersection between mapping, urbanism, and strategic design driven by the capacity of cartographic pieces to project data over different landscapes. Tomateo holds a Master's in Urban Design from Harvard University Graduate School of Design and a Bachelor's in Architecture from Pontificia Universidad Católica del Perú.

Timon McPhearson is Director of the Urban Systems Lab and Associate Professor of Urban Ecology at The New School. He is a Research Fellow at The Cary Institute of Ecosystem Studies and Stockholm Resilience Centre and a member of the IPCC.

Chapter 11
Anticipatory Resilience Bringing Back the Future into Urban Planning and Knowledge Systems

Tischa A. Muñoz-Erickson, Kaethe Selkirk, Robert Hobbins, Clark Miller, Mathieu Feagan, David M. Iwaniec, Thaddeus R. Miller, and Elizabeth M. Cook

Abstract Anticipatory thinking is a critical component in urban planning practices and knowledge systems in an era of unpredictability and conflicting expectations of the future. This chapter introduces "anticipatory resilience" as a futures-oriented knowledge system that intentionally addresses uncertain climate conditions and explores alternative, desirable future states. It suggests a portfolio of tools suitable for building long-term foresight capacity in urban planning. Examples of knowledge systems interventions are presented to explore the trade-offs, constraints, possibilities, and desires of diverse future scenarios co-generated in settings with people that hold different perspectives, knowledge, and expectations.

Keywords Resilient futures · Climate change · Urban planning · Anticipatory capacities · Knowledge systems

---

T. A. Muñoz-Erickson (✉)
USDA Forest Service, International Institute of Tropical Forestry, Río Piedras, PR, USA
e-mail: tischa.a.munoz-erickson@usda.gov

K. Selkirk · C. Miller · M. Feagan
School for the Future of Innovation in Society, Arizona State University, Tempe, AZ, USA

R. Hobbins
School of Sustainability, Arizona State University, Tempe, AZ, USA

D. M. Iwaniec
Andrew Young School of Policy Studies, Urban Studies Institute, Georgia State University, Georgia, USA

T. R. Miller
School of Public Policy, University of Massachusetts Amherst, Amherst, USA

E. M. Cook
Department of Environmental Science, Barnard College, NY New York, USA

Z. A. Hamstead et al. (eds.), *Resilient Urban Futures*, The Urban Book Series,
https://doi.org/10.1007/978-3-030-63131-4_11

11.1 Introduction

Thinking about the future of cities is the central focus of this book. Rapidly changing social, technological, environmental, and climate conditions pose unique challenges to the way urban planners, decision-makers, designers, and citizens think, plan ahead, and take actions to build cities that are resilient to future change. Resilience and disaster scholars alike expect governments, institutions, and civic organizations to anticipate future risks and the occurrence of shocks and stresses in a proactive manner to mitigate and adapt (Baud and Hordijk 2009; Aguirre 2006). Yet, although anticipation is considered an important component of both urban planning and resilience, the concept in both fields would benefit from moving beyond a bias toward quantitative predictive modeling, and toward the capacity building practices that allow different actors in the city to engage with planning long-term resilient futures (Myers and Kitsuse 2000; Boyd et al. 2015).

In this chapter, we are concerned with making anticipation a central practice in adaptation and resilience research, planning, and action in cities. Indeed, cities have a long history of imagining and planning for the future. Since the times of Kevin Lynch and Le Corbusier, urban planners and designers have conjured up different ideas and images of what cities can and should look like. The very purpose of planning is to prepare for future activity (Myers and Kitsuse 2000). We question, however, the extent to which the knowledge systems currently employed by planners to think about the future are capable of building anticipatory resilience. Knowledge systems are the social and institutional practices, tools, and norms that organizations use to generate, validate, circulate, and use knowledge in decision-making, policy, and design (Miller and Muñoz-Erickson 2018). Cities need to become more ambitious about how they factor the high unpredictability and uncertainty of climate change into their knowledge systems. City planners and policy-makers need a more effective future-oriented approach that enables them to comprehend present and future complexity (Ratcliffe and Krawczyke 2011).

We promote a systematic and rigorous exploration, understanding, and imagining of plausible and desirable futures that enable cities to consider the wide-reaching implications of design policy and planning choices. Anticipatory resilience is also crucial in disaster planning if communities and governments are seeking not to just "bounce back" after a shock or disaster, but to use that moment as a 'window of opportunity' to transform urban communities along more sustainable pathways (Eakin et al. 2018). Thus, organizations must rethink their knowledge systems (Feagan et al. 2019; Muñoz-Erickson et al. 2017) to anticipate impending change and shape a preferred future condition. To help cities address this challenge, we draw upon the ideas, practices, and techniques used in the fields of anticipation and futures research to suggest ways that knowledge systems integrate foresight.

We begin with a discussion on why climate uncertainties and complexities pose such a difficult challenge to urban planning and why the traditional risk-based knowledge systems are not well suited to handle these uncertainties. We then present the main argument for a more sophisticated and ambitious definition of anticipation to help planners think about the future of cities differently in the face of deep uncertainties. Contrary to the near-term future that the urban planning field typically works with, the future that anticipatory resilience deals with is nonimmediate (far enough away to be deeply uncertain), opening a big space for a variety of actors to put their differences on the table and collectively come up with a vision to act on (Alvial-Plavicino 2015). We finish by showcasing a portfolio of foresight methods for designing future-based knowledge systems capable of building anticipatory resilience, using examples from the Urban Resilience to Extremes Sustainability Research Network (UREx SRN). The result is a strengthened ability for practitioners and communities to explore, deliberate, and steer future pathways while embracing the uncertainties associated with climate change. In other words, against tendencies to "wait and see" or leave the long-term future in the hands of the biggest players, we put forward an approach that builds the capacity of practitioners and researchers from various sectors to come together and talk about the visions we need today to create the policy frameworks, knowledge systems, and governance relationships for a resilient tomorrow.

11.2 The Challenge of Deep Climate Uncertainty

Unlike risks, which can be reduced by quantifying the "likelihood" or probability, "uncertainty" is a state of knowledge where probabilities or likelihoods cannot be confidently defined and quantified (Stirling and Scoones 2009). Climate change uncertainties go beyond trends and changes in future atmospheric conditions. New forms of uncertainties around local capacities to respond to climate change and the effectiveness of responses, including changes in human behaviors, also pose a challenge to urban planning. Stults and Larsen (2018) recently reviewed climate adaptation planning literature and identified thirteen types of climate-related uncertainties that local city planners are facing. These uncertainties were grouped into four categories, including: (1) uncertainty in future climate conditions; (2) uncertainty in climate-related behaviors and political decisions external to the municipality; (3) uncertainty in climate-related local coping capacity; and (4) uncertainty in effective local responses. While most scientific work efforts have focused on reducing sources of uncertainty in future climate conditions and climate-related behaviors and political decisions, these uncertainties are outside a city planner's direct influence. Therefore, city planning efforts to apply uncertainty reducing-techniques for these uncertainties will be fruitless since they fall outside the local solution space.

On the other hand, although coping capacities and effectiveness of responses (Categories 3 and 4 above) fall under the direct influence of local planners, there are also large knowledge gaps that further complicate these types of uncertainties (Stults and Larsen 2018). With respect to, for instance, our urban infrastructure—roads, buildings, water, power, etc.—there is unpredictability in the extent to which it can adapt to accelerated climate change. This is because of the decades and centuries that our infrastructure has had to withstand the building of interconnecting infrastructure, embedding of new hardware, and most recently, implementation of new technologies (e.g., sensors and computing, automation) (Miller et al. 2018; Chester and Allenby 2018). Similarly, knowing the conditions that enable communities to effectively cope with changing climate conditions is difficult to ascertain (Stults and Larsen 2018), especially when many of the analytical approaches used to evaluate community vulnerability and adaptive capacities are limited to static, place-based attributes and miss other important, dynamic dimensions of coping capacities (Eakin et al. 2018).

Finally, the uncertainty surrounding the climate change discourse in the adversarial American political arena is further compounded by resilience as a concept that also engages normative dimensions in urban planning. In other words, how *should* we develop resilient cities? There is an extensive debate in the resilience academic literature as to whether the concept of resilience should be used as a descriptive concept of system change, or whether it is a normative concept because of the power dynamics that shape resilience policies and outcomes (Brand and Jax 2007; Olsson et al. 2015). We take the position here that any application of resilience in practice will be political and involve negotiations of diverse actors and interests on what are desirable and preferred pathways of transformation (Harris et al. 2018). It is precisely because anticipation deals intentionally with the normative dimensions of envisioning the future—the expectations, values, imaginations, desires—of society collectively that it offers a powerful framework for resilience planners to "open-up" and engage with the politics of resilience as they plan for the future. Applying strategic foresight with aspirational tools offers a way to ask the "resilience of what, to what, and for whom?" that many resilience scholars are asking for (Meerow and Newell 2016).

Exploring the politics of urban resilience with foresight also facilitates the leveraging of postdisaster reconstruction stages to build long-term transition pathways toward sustainability-oriented visions (Brundiers and Eakin 2018). Sustainability scholars argue that having a negotiated and articulated vision of an alternative development pathway prior to an event will make it possible for willing actors to take advantage of "windows of opportunity" after disasters and carry forward the ideas and strategies, even in the midst of significant hardship and loss (Eakin et al. 2018). On the other hand, the absence of transformative visions prior to an event usually results in powerful interests taking advantage of postdisaster recovery to further the status quo in the name of "building back better."

11.3 Limits of Risk-Based Knowledge Systems

How planners deal with risk and uncertainty is a crucial differentiating factor between planning practices and the forward-looking approaches required to address climate change. Although planners, both in academic and professional circles, have had a special relationship with the future (Myers and Kitsuse 2000), in practice, the dominant approaches to exploring the future have been tools for projections and forecasts to acquire predictive knowledge about the future, or as Quay (2010) calls it, the "predict and plan" approach. *Projections* are described as accounting systems that rely on hypothetical assumptions of the past and then expect or project the same trends or behaviors to continue into the future. They usually contain conditional terms such as "if/then" statements about the future (Myers and Kitsuse 2000). While they are not technically predictive, planners often mistake them as such (Isserman 1984; Myers and Kitsuse 2000).

Forecasts, on the other hand, are predictive and provide planners with a likelihood about a future state or behavior derived using statistical calculations and models (Isserman 1984; Myers and Kitsuse 2000). The best example is a weather forecast that uses observable, quantitative data to characterize current weather conditions and predict future atmospheric conditions with computer models. Because forecasts are based on a model, the quality of their results represent a best guess about the future and depend on the assumptions and the input data that were used in that model. With respect to physical urban planning, including land use, transportation, and water infrastructure, forecasts are used to examine trends over time or a desired future state and then design the infrastructure needed to serve that future.

Tools like projections and forecasts are risk-based knowledge systems that underlie the "predict and plan" approach and are not sufficient to address the conditions of deep climate uncertainty. As Selkirk et al. (2018) explain, while linear and quantitative modes of knowing the future are useful in a wide range of settings, "they structure our engagement with the future down to a limited number of model runs, numbers, or decimal points" (p. 6). The future can be more complex and dynamic than numbers fully account for (Ibid). Different from the "predict and plan" approach, anticipatory resilience recognizes that some aspects of the future are unknowable and different from the present, and therefore a systemic understanding of how multiple trends and visions will extend forward and interact with one another is useful to shape new possibilities and patterns of behavior in the process.

11.4 Toward More Anticipatory Resilience

Anticipation is an act of looking toward the future, or being forward-looking. At an individual level, we may think of anticipation as "knowing what is coming" or "getting ahead of ourselves." Expectations play a central role in anticipation because how we come to know the future (e.g., tools, values, cultures, etc.) guides what we expect from it, and in turn, helps to shape present and future action (Selkirk et al. 2018; Selin 2008). Expectations, then, are key in understanding, building, and enacting anticipatory capacity (Alvial-Palavicino 2015). Yet, understanding expectations of the future is only part of anticipation. To act in an anticipatory way means to act in relation to the future—and knowledge of emerging transformations—such that what we expect of the future changes our decisions or behaviors today (Polasky et al. 2011). As Poli (2017) puts it, a weather forecast in itself is not anticipatory, but taking an umbrella as a consequence of watching the weather forecast is an anticipatory behavior.

At a societal or collective level, anticipation means that the "future" is made actionable by a set of societal arrangements, attitudes, and interventions (Alvial-Palavicino 2015). Commonly discussed in the literature as anticipatory capacity or anticipatory governance, anticipation refers to a model of decision-making under very high uncertainty (Quay 2010). Scholars of emerging technologies and responsible innovation define anticipation as the ability to rehearse future possibilities prior to "diving into the future" to help steer technology and development towards socially desirable situations (Guston 2014; cited in Alvial Palavicino 2015). The field of sustainability transitions anticipates long-term visions to develop transition pathways and actions toward those visions (Boy et al. 2015; Wiek and Iwaniec 2014). Anticipation is thus concerned with extended time horizons, where the future is open-ended and unpredictable. Building foresight capacity—or what some describe as "futures literacy" (Larsen et al. 2020)—is a key goal in anticipation, allowing us to imagine alternative futures and test courses of action before we deploy them (Fuerth 2009; Wachs 2001).

Visioning and scenario building efforts have tried to gain traction in the planning field in recent years, but the absence of specific strategies for achieving goals and the inability of these efforts to become anything but wish lists for the future has received much criticism, citing them as shortsighted and hollow (Myers and Kitsuse 2000). In their review of 44 US local climate adaptation plans, Stults and Larsen (2018) found that none of these plans used scenario planning or other techniques to explore the future. This finding confirms the observation made by Myers and Kitsuse (2000) that the field of planning has lost sight of the future, despite its future-oriented characterization in the literature (Myers and Kitsuse 2000).

Anticipatory resilience uses tools and practices that enable long-term foresight planning. This approach explicitly calls for reflexivity, or the self-awareness to reveal the assumptions and intentions one makes about what the future will look like, to clearly articulate and negotiate the politics and unintended consequences that are embedded in creating alternative futures, and to recognize when changes in our knowledge systems or actions are necessary to steer away from maladaptive or unjust outcomes. Adaptability through monitoring, feedback, and learning are therefore key elements of this anticipatory approach (Boyd et al. 2015).

11.4.1 Portfolio of Future-Based Knowledge Systems

Bengston (2019) and Stirling (2004) have reviewed a variety of methods and techniques from the field of future studies that are relevant to our discussion. In Fig. 11.1, we show common future methods and techniques in relation to their utility for resilience planners to engage with uncertainty, time horizons, and dimensions of the future. Exploratory tools represent those used to know and articulate the future in a more "open" way, based on visions, aspirations, and expectations, rather than just on what the data tells us could happen. The most common tools are qualitative

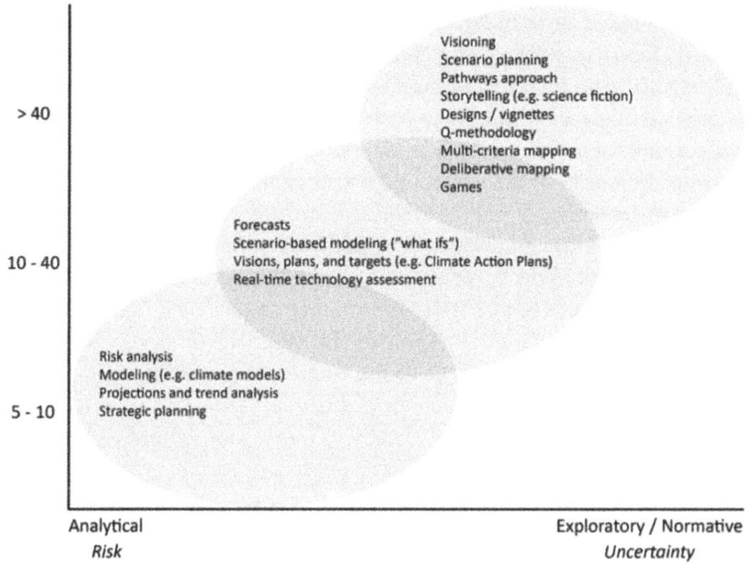

Fig. 11.1 Future methods and techniques and their utility in engaging with uncertainty, time horizons, and dimensions of the future

and include visioning and scenario planning. As a practice to represent and evoke a shared preferred image of the future to guide action, *visioning* has been used in the development of master and general plans for cities. A commonly cited example is the Atlanta's Vision 2020, a regional visioning effort that the city carried out in the early 1990s. Visioning has also been used as a way to encourage citizen involvement in a collaborative process toward shared understanding and optimistic picture of a plausible future (Myers and Kitsuse 2000).

Scenario planning is the practice that helps give action to these visions by aiding planners in developing narratives or stories to specify the sequence of actions and events that impact planning decisions and lead to the desirable state, or vision (Myers and Kitsuse 2000). Scenario planning is the most well-known future practice, having emerged from the military in the 1950s and been widely applied in business corporations by the 1970s as a form of strategic planning (Bengston 2019). Scenario planning allows input of qualitative measures into quantitative forecasts and merges technical and participatory planning to help address uncertainty in creative ways (Chakarborty and McMillan 2015). Visioning and scenario planning are therefore not meant to be predictive, but to instead allow qualitative and quantitative modes of knowing to come together and mutually inform each other. The pathways approach is another useful method employed by the climate adaptation community to support decision-makers and communities in envisioning alternative scenario pathways, which are to be met through a sequence of adaptation actions and triggers that are managed and monitored over time (Wise et al. 2014; Barnett et al. 2014). Scenario and pathway approaches are useful analytical techniques that support the exploration of a variety of uncertainties and long-term horizons and connect them with specific short-term actions in a dynamic way (Haasnoot et al. 2013).

In addition to visioning and scenario planning, there are a number of future research methods that use more imaginative or creative techniques to foster "out-of-the-box" thinking when exploring potential futures. *Games* are participatory and creative techniques—including cards, board games, role-playing exercises, and online games—for active learning, creating foresight, and problem-solving. The Urban Sustainability Directors Network (USDN) uses the "Game of Floods" to help city practitioners think about a variety of planning and action scenarios to address flood risks in their cities (Baja 2019). *Storytelling* and science fiction are methods for creative and imaginative exploration of the future. Stories can describe plausible futures and connect the present to the future using narratives that link cause and effect and illustrate the consequences of key events, decisions, or technological innovations. Miller et al. (2015), for instance, found storytelling to be a valuable method to open deliberation and scenario development to a diverse group of energy and nonenergy professionals about the future of energy in Arizona in 2050.

Because exploring potential futures is an open-ended activity, visions and scenarios should be generated and deliberated using *participatory frameworks* so that the futures are co-produced, and inclusive of multiple voices, perspectives, and knowledge systems. In addition to participatory action methods, Stirling (2004) suggests a number of decision-analytic techniques that facilitate the evaluation of trade-offs among multiple values and discourses, including "Q methodology," "multicriteria mapping," and "deliberative mapping." Quantitative scenario-based modeling is sometimes used in combination with the scenario narratives to explore outcomes or trade-offs of the strategies and interventions that are part of the narrative. The methods and techniques presented all have their strengths and limitations. Therefore, instead of adhering to one single approach or tool to explore the future, we recommend a portfolio that includes a variety of tools and methods to explore futures.

11.5 Examples of Knowledge Systems Interventions to Build Anticipatory Resilience

We made knowledge systems innovations toward building anticipatory resilience as part of the Urban Resilience to Extremes Sustainability Research Network (UREx SRN; https://URExSRN.net) in three ways. One is through a scenario development approach that we carried out in nine cities in the USA and Latin America to articulate and explore the implications of positive futures for urban resilience. As we describe in greater detail in Chap. 6, this approach begins with an analysis of existing governance framings, perspectives, knowledge, and values that different actors, including government, civil sector organizations, academia, and private sector groups, have with respect to climate resilience and the future of the city. Along with other assessments of existing social, ecological, and technological conditions [e.g., existing municipal strategies (Chap. 3) and vulnerability analysis (Chap. 4)] these were used in a co-production process with local researchers and practitioners to define a set of climate and urban challenges (e.g., extreme heat) and themes (e.g., energy security), as well as to identify a diverse set of stakeholders to work on these context-based scenarios for their city's future (Chap. 7). During the participatory process, participants worked in small groups with trained facilitators to collectively define a vision and goals for a very long-term horizon—all the way out to the year 2080. Through a series of structured activities, participants also defined short- and mid-term actions, the specific locations where these actions would need to take place, and the linkages between strategies necessary to realize their long-term vision.

The combination of analytical and exploratory techniques and activities we used to guide the co-production of positive futures over very long-time frames (to 2080) helped "open up" discussions about the uncertainties and challenges that cities face, while allowing participants who do not normally work together to think "outside the box" about what very transformative strategies might entail, including social equity outcomes. We used activities designed to stress test the scenarios (e.g., disaster cards) and trigger changes in actions that could lead to maladaptive outcomes. Opening up the future through this structured process allowed participants to navigate uncertainties and different values in a safe space where differences were encouraged to spur innovative ideas. Not surprisingly, navigating these value differences was often a challenge and deliberations sometimes got very heated. These conflicts were often about short-term barriers posed by the current system (e.g., zoning code regulations), so when participants were reminded that the very long-time frame being discussed allowed for transformative thinking, their perceptions shifted again towards common values and the creative innovations needed to move forward in radical ways. In the end, some of the scenario interventions were successful in producing future visions that became guides for short-term actions, while some stayed at the discussion level. Nevertheless, for a number of our UREx cities, the scenario process served as an archetype for how to plan using anticipatory practices and opened up a new space to negotiate the various values and meanings of a resilient city.

Our second innovative knowledge system intervention to build anticipatory resilience was the UREx SRN Resilient Coastal Cities (RC2) Innovation Labs, where we engaged city practitioners, neighborhood residents, NGO leaders, resilience researchers, engineers, and data visualization specialists in Miami, San Juan, and Baltimore to co-design an integrated data visualization platform. The goal of the Innovations Labs was to help increase anticipatory capacities through access, use, and sharing of information and data on resilience to coastal climate risks. With support from the National Science Foundation's Smart and Connected Communities program, these Innovation Labs served as spaces for participants to evaluate the suitability, relevance, and quality of different data visualization tools with respect to the various knowledge systems practices of their organizations, such as by developing reports to meet municipal code standards or to explore different sea-level

Fig. 11.2 Innovation Lab activities in the community of Santurce in San Juan, Puerto Rico. Participants classified different types of data and visualizations according to their relevance to different use cases, including climate education, implementation of adaptation strategies, or development of future coastal scenarios (top right). Participants also created an actor map of the various governmental and nongovernmental organizations and institutions carrying out climate adaptation and resilience initiatives in Santurce following Hurricane María in 2017 (left and bottom right)

rise scenarios (Fig. 11.2). The Labs resulted in a prototype of what a smart knowledge system could look like, a system by which local needs are prioritized through community empowerment. Such a system would allow for the evaluation of existing vulnerabilities and the anticipation of potential futures by employing data visualizations and connecting different governance sectors, communities, and knowledge systems across the three Atlantic coastal cities.

A final intervention was a Resilience Governance Workshop in Portland, Oregon. The UREx team had worked closely with City of Portland practitioners and identified organizational barriers and the need for innovation in governance as critical areas for resilience. The workshop was designed collaboratively with practitioner partners and focused on building transformative governance principles into resilience governance proposals generated in the workshop. The workshop produced four proposals for resilience governance structures. Based on the exercises, groups were challenged to integrate issues of foresight and anticipation, including learning and experimentation, diversity of communities and knowledge types, and the ability to identify and unlock path dependencies and mal-adaptations in terms of how organizations think about the future and uncertainties. The results from this workshop will be further developed and incorporated into resilience planning in the City of Portland.

11.6 Conclusion

Developing transformative pathways for sustainable and resilient cities hinges on the ability of city officials, policy-makers, businesses, scientists, civic leaders, and residents to think, know, and decide on future strategies in an era of unpredictability and conflicting expectations of the future. True resilience can only result from genuinely transformative ideas, policies, and practices concerning how societies go about reducing risk (Tierney 2014). Although planning for the future is at the core of urban planning, current risk-based knowledge systems that rely on predictive approaches are not enough to address the complexities and uncertainties that climate change brings for cities. Anticipation is a critical component of building resilience but needs to be better embedded in urban planning practices and knowledge systems. We have argued for an anticipatory resilience approach to future-based knowledge systems that intentionally explores alternative desirable future states and have offered suggestions for a portfolio of tools suitable to building long-term foresight capacity in urban planning, including scenario planning, games, storytelling, and multicriteria mapping, to name a few. We have presented three examples from the UREx SRN of knowledge systems interventions where we used a combination of foresight tools that resulted in multiple urban futures, transformative governance structures, as well as an integrated data visualization platform to help explore these futures. We contend that future-based knowledge systems are suitable to explore the trade-offs, constraints, possibilities, and desires of different future scenarios co-generated in settings with people from different perspectives, knowledge, and expectations.

References

Aguirre BE (2007) Dialectics of vulnerability and resilience. Georgetown J Poverty Law Policy 14(1):39–59

Alvial-Palavicino C (2016) The future as practice. A framework to understand anticipation in science and technology. Tecnoscienza: Ital J Sci Technol Stud 6(2):135–172

Baja K (2019) Game of floods. Urban Sustainability Directors Network. https://www.usdn.org. Accessed 15 Jan 2020

Barnett J, Graham S, Mortreux C et al (2014) A local coastal adaptation pathway. Nat Clim Change 4(12):1103–1108. https://doi.org/10.1038/nclimate2383

Baud ISA, Hordijk MA (2009) Dealing with risks in urban governance: What can we learn from 'resilience thinking.' In: Paper presented at the 4th International Conference of the International Forum on Urbanism (IFoU), Amsterdam/Delft

Bengston DN (2019) Futures research methods and applications in natural resources. Soc Nat Resour 32(10):1099–1113. https://doi.org/10.1080/08941920.2018.1547852

Boyd E, Nykvist B, Borgström S et al (2015) Anticipatory governance for social-ecological resilience. Ambio 44(Suppl 1):S149–S161. https://doi.org/10.1007/s13280-014-0604-x

Brand FS, Jax K (2007) Focusing the meaning(s) of resilience: resilience as a descriptive concept and a boundary object. Ecol Soc 12(1):23

Brundiers K, Eakin HC (2018) Leveraging post-disaster windows of opportunities for change towards sustainability: a framework. Sustain Sci Pract Pol 10(5):1390. https://doi.org/10.3390/su10051390

Chakraborty A, McMillan A (2015) Scenario planning for urban planners: toward a practitioner's guide. J Am Plann Assoc 81(1):18–29. https://doi.org/10.1080/01944363.2015.1038576
Chester MV, Allenby B (2018) Toward adaptive infrastructure: flexibility and agility in a non-stationarity age. Sustain Resil Infras 4(4):173–191. https://doi.org/10.1080/23789689.2017.1416846
Eakin H, Muñoz-Erickson TA, Lemos MC (2018) Critical lines of action for vulnerability and resilience research and practice: lessons from the 2017 hurricane season. J Extreme Events 05(02n03):1850015. https://doi.org/https://doi.org/10.1142/S234573761850015X
Feagan M, Matsler M, Meerow S et al (2019) Redesigning knowledge systems for urban resilience . Environ Sci Policy 101:358–363. https://doi.org/10.1016/j.envsci.2019.07.014
Fuerth LS (2009) Foresight and anticipatory governance. Foresight 11(4):14–32. https://doi.org/10.1108/14636680910982412
Guston DH (2014) Understanding "anticipatory governance." Social Stud Sci 44(2):218–242. https://doi.org/10.1177/0306312713508669
Haasnoot M, Kwakkel JH, Walker WE et al (2013) Dynamic adaptive policy pathways: a method for crafting robust decisions for a deeply uncertain world. Global Environ Change: Hum Policy Dimens 23(2):485–498. https://doi.org/10.1016/jgloenvcha.2012.12.006
Harris LM, Chu EK, Ziervogel G (2018) Negotiated resilience. Resilience 6(3):196–214. https://doi.org/https://doi.org/10.1080/21693293.2017.1353196
Isserman AM (1984) Projection, forecast, and plan on the future of population forecasting. J Am Plann Assoc 50(2):208–221. https://doi.org/10.1080/01944368408977176
Ratcliffe J, Krawczyk E (2011) Imagineering city futures: the use of prospective through scenarios in urban planning. Futures 43(7):642–653. https://doi.org/10.1016/j.futures.2011.05.005
Larsen N, Kaeseler Mortensen J, Miller R (2020) What is 'futures literacy' and why is it important? Medium.https://medium.com/copenhagen-institute-for-futures-studies/what-is-futures-literacy-and-why-is-it-important-a27f24b983d8. Accessed 12 Feb 2020
Meerow S, Newell JP (2016) Urban resilience for whom, what, when, where, and why? Urban Geogr 40(3):1–21. https://doi.org/10.1080/02723638.2016.1206395
Miller CA, O'Leary J, Graffy E et al (2015) Narrative futures and the governance of energy transitions. Futures 70:65–74. https://doi.org/10.1016/j.futures.2014.12.001
Miller T, Chester M, Muñoz-Erickson TA (2018) Rethinking infrastructure: resilience in an era of unprecedented events. Iss Sci Technol(Winter):45–58
Miller CA, Muñoz-Erickson TA (2018) The rightful place of science: designing knowledge. Consortium for Science, Policy & Outcomes, Tempe, Arizona
Muñoz-Erickson TA, Miller CA, Miller TR (2017) How cities think: knowledge co-production for urban sustainability and resilience . For Trees Livelihoods 8(6):203. https://doi.org/10.3390/f8060203
Myers D, Kitsuse A (2000) Constructing the future in planning: a survey of theories and tools. J Plann Educ Res 19(3):221–231. https://doi.org/10.1177/0739456X0001900301
Olsson L, Jerneck A, Thoren H et al (2015) Why resilience is unappealing to social science: theoretical and empirical investigations of the scientific use of resilience. Sci Adv 1(4):e1400217–e1400217. https://doi.org/10.1126/sciadv.1400217
Polasky S, Carpenter S, Folke C, Keeler B (2011) Decision-making under great uncertainty: environmental management in an era of global change. Trends Ecol Evol 26(8):398–404. https://doi.org/10.1016/j.tree.2011.04.007
Quay R (2010) Anticipatory governance: a tool for climate change adaptation. J Am Plann Association 76(4):496–511. https://doi.org/10.1080/01944363.2010.508428
Selin C (2008) Sociology of the future: tracing stories of technology and time. Socio Compass 2(60):1875–1895. https://doi.org/10.1111/j.1751-9020.2008.00147
Selkirk K, Selin C, Felt U (2018) A festival of futures: recognizing and reckoning temporal complexity in foresight. Handbook of anticipation: Theoretical and applied aspects of the use of future in decision making. Springer, Switzerland

Stirling A, Scoones I (2009) From risk assessment to knowledge mapping: science, precaution, and participation in disease ecology. Ecol Soc 14(2):14. https://doi.org/https://doi.org/10.5751/ES-02980-140214

Stults M, Larsen L (2018) Tackling uncertainty in US local climate adaptation planning. J Plann Educ Res 00:1–16. https://doi.org/10.1177/0739456X18769134

Tierney K (2014) The social roots of risk: producing disasters, promoting resilience. Stanford University Press, Stanford, California

Wachs M (2001) Forecasting versus envisioning: a new window on the future. J Am Plann Assoc 67(4):367–372. https://doi.org/10.1080/01944360108976245

Wiek A, Iwaniec D (2014) Quality criteria for visions and visioning in sustainability science. Sustainability Sci 9(4):497–512. https://doi.org/10.1007/s11625-013-0208-6

Wise RM, Fazey I, Stafford Smith M et al (2014) Reconceptualising adaptation to climate change as part of pathways of change and response. Global Environ Change: Hum Policy Dimens 28:325–336. https://doi.org/10.1016/j.gloenvcha.2013.12.002

Tischa A. Muñoz-Erickson is a Research Social Scientist in the USDA Forest Service's International Institute of Tropical Forestry, in Río Piedras, Puerto Rico. She studies urban sustainability governance, including the policy networks, knowledge systems, anticipatory capacities, and strategies to advance sustainability, resilience, and equity. She is also actively involved in transdisciplinary platforms to facilitate the co-production of futures and transition pathways in the U.S. and Latin American cities. Muñoz-Erickson leads the International Urban Field Station and the San Juan ULTRA and is also Co-PI of the National Science Foundation Urban Resilience to Extreme Events Sustainability Research Network (UREx SRN) and the SETS Convergence project.

Kaethe Selkir holds a Ph.D. in Science and Technology from Arizona State University and master's degrees in Community and Regional Planning and Sustainable Design from the University of Texas at Austin School of Architecture. Her work examines the tools and processes that organizations use to make decisions about the future.

Robert Hobbins is a Ph.D. Candidate at the School of Sustainability at Arizona State University and an NSF Research Intern at the USDA Forest Service International Institute of Tropical Forestry.. Robert has a BS in Physics and a Master of Education from The Ohio State University. He also earned an MS in Community Resources & Development–Sustainable Communities at Arizona State University. Robert joined the Urban Resilience to Extremes Sustainability Research Network (UREx SRN) as a Graduate Fellow in 2016 and continues to be active in the network by supporting the planning and implementation of scenario workshops and working with UREx SRN city practitioners to build local capacities to transition to more sustainable and resilient urban futures.

Clark Miller is Professor and Director of the Center for Energy and Society in the School for the Future of Innovation in Society at Arizona State University. His work offers theoretical, practical, and critical insights into how societies can better design and deploy knowledge and technology in the service of more just and sustainable futures

Mathieu Feagan is a critical interdisciplinary scholar whose work focuses on knowledge mobilization for climate justice across diverse communities of professionals, activists, and researchers. Through innovative pedagogical practices and action research projects across the U.S., Latin America, and Canada, Matt is building new capacity for challenging dominant power relations and shifting institutional norms toward deeper forms of democracy, ecological consciousness, and human liberation. He is based in the School for the Future of Innovation in Society at Arizona State University, in Tempe, Arizona.

David M. Iwaniec is an Assistant Professor of Urban Sustainability at the Urban Studies Institute, Andrew Young School of Policy Studies at Georgia State University. He is a sustainability scientist researching anticipatory and systems approaches to advance urban sustainability, resilience, and justice. His work focuses on the co-development of scenarios and transition pathways for positive futures of urban transformation.

Thaddeus Miller is an Associate Professor at the School of Public Policy at the University of Massachusetts Amherst. He collaborates with interdisciplinary teams of researchers and practitioners to enable cities to leverage science and technology to meet policy goals and community needs. He is on the Executive Management Team for the National Science Foundation funded Urban Resilience to Extremes Sustainability Research Network, and co-PI of the NSF-funded STIR Cities project. His recent book, Reconstructing Sustainability Science: Knowledge and Action for a Sustainable Future, part of the Earthscan Routledge Science in Society Series, examines how scientists can navigate epistemic and normative ten-sions to link knowledge to social action. Prior to joining UMass, Dr. Miller was a faculty member at Portland State University and Arizona State University

Elizabeth M. Cook is an Assistant Professor at Barnard College in the Department of Environmental Science. She is an urban ecosystem ecologist and her research focuses on future urban sustainability and human-environment feedbacks in urban and nearby native ecosystems. She conducts research on sustainability and resilience planning through participatory scenario development with local stakeholders. Her work seeks to understand cities as social-ecological-technological systems with a comparative approach in Latin American and U.S. cities. Cook holds a Ph.D. from Arizona State University and a Bachelor of Arts from Wellesley College.

Chapter 12
A Vision for Resilient Urban Futures

Timon McPhearson, David M. Iwaniec, Zoé A. Hamstead, Marta Berbés-Blázquez, Elizabeth M. Cook, Tischa A. Muñoz-Erickson, Lelani Mannetti, and Nancy Grimm

Abstract A fundamental systems approach is essential to advancing our understanding of how to address critical challenges caused by the intersection of urbanization and climate change. The social–ecological–technological systems (SETS) conceptual framework brings forward a systems perspective that considers the reality of cities as complex systems and provides a baseline for developing a science of, and practice for, cities. Given the urgency of issues we collectively face to improve livability, justice, sustainability, and resilience in cities, bringing a systems approach to resilience planning and policymaking is critical, as is development of positive visions and scenarios that can provide more realistic and systemic solutions. We provide a vision for more resilient urban futures that learns from coproduced scenario

T. McPhearson (✉)
Urban Systems Lab, The New School, New York, NY, USA
e-mail: timon.mcphearson@newschool.edu

Cary Institute of Ecosystem Studies, Millbrook, NY, USA

Stockholm Resilience Center, Stockholm University, Stockholm, Sweden

D. M. Iwaniec · L. Mannetti
Urban Studies Institute, Andrew Young School of Policy Studies, Georgia State University, Georgia, USA

Z. A. Hamstead
Department of Urban and Regional Planning, School of Architecture and Planning, University At Buffalo, State University of New York, Buffalo, NY, USA

M. Berbés-Blázquez
School for the Future of Innovation in Society, Arizona State University, Tempe, AZ, USA

E. M. Cook
Department of Environmental Science, Barnard College, New York, NY, USA

T. A. Muñoz-Erickson
USDA Forest Service, International Institute of Tropical Forestry, Rio Piedras, PR, USA

N. Grimm
School of Life Sciences, Arizona State University, Arizona, USA

Julie Ann Wrigley Global Institute of Sustainability, Arizona State University, Tempe, AZ, USA

Z. A. Hamstead et al. (eds.), *Resilient Urban Futures*, The Urban Book Series,
https://doi.org/10.1007/978-3-030-63131-4_12

development work in nine US and Latin American cities in the URExSRN. We find that developing an urban systems science that can provide actionable knowledge for decision-making is an emerging, and much needed, transdisciplinary research agenda. It will require true boundary-crossing to bring the knowledge, skills, tools, and ideas together in ways that can help achieve the normative goals and visions we have for our shared urban future.

Keywords Anticipatory resilience · Co-production · Social-ecological-technological systems (SETS) · Urban systems science · Visioning · Urban futures

12.1 Bringing Positive Futures into Research and Practice

Much of the discourse around urban and global futures tends to be dystopian, featuring visions of environmental and societal collapse along with business-as-usual forecasts. Negative outlooks, scenarios, and projections across mainstream press, political statements, and academic literature are abundant in popular narratives and, for most of us, dominate the discourse about our future (Bennett et al. 2016). Dystopian forecasts at both local and global levels make it difficult to develop actionable plans and policies for generating more positive futures. Despite the role of cities as global nodes of transformational innovation, such dystopian discourses pervade the discussion and visions of the future of cities. Although the dramatic speed and scale of urbanization is driving the articulation of the twenty-first century as the time when we will live on an urban planet (Elmqvist et al. 2019), it is also recognized as a time bringing about both perverse challenges as well as critical opportunities for fundamental transformations of how we build, design, plan, and govern our cities.

Instead of dystopian visions, we need to imagine and coproduce (see Chap. 7) shared positive visions that can support the development of more transformative plans, policies, and actions to guide our decisions now toward the building of a longer term, more just, more resilient, and more sustainable world. In short, positive visions are critical to guide urban planning, motivate actions, inspire innovative strategies, and move toward transformative change. Negative discourse around urban futures often leaves little reason to invest in long-term social and environmental goods, nor does it recognize the vast possibility within existing creativity and innovation that already drives much of development in cities (McPhearson et al. 2017).

In the Urban Resilience to Extremes Sustainability Research Network (URExSRN), we showcase how researchers and practitioners can come together to craft visions of resilient urban futures. These positive futures are intended to explore radical departures of the status quo—when small tweaks are not enough. The entire process is about creating opportunities for stakeholders to anticipate, imagine, and scrutinize the plausible pathways to desirable futures. The focus on positive futures in URExSRN was realized primarily through a scenario coproduction process conducted in each of the nine network cities (Chaps. 6 and 7). Key learnings from conducting this work in nine US and Latin American cities show (a) the urban social–ecological–technological systems (SETS) context that both present-day patterns and processes, as well as future visions, plans, and scenarios need to examine and work within, (b) a need to understand the structural inequities built into people's present experiences and histories, (c) how both science and practice need to advance systems approaches together as part of visioning work, and (d) the critical need for practice to put goals and strategies into action for transformation.

12.2 Thinking in Systems

Cities face multiple risks that can overlap in space and time. For example, immediately following Hurricane Sandy that made landfall on the East Coast of the USA in September 2012, a cold front with Arctic air blew into the region and created severe heating challenges in a region where Sandy left millions without power and thus no access to heat. In 2020, policies intended to decrease health and mortality impacts of the COVID-19 pandemic in New York City induced a radical shift in the locations where people needed services (primarily in their homes and apartments), affecting demand for energy, transit, and green spaces. This shift, in turn, created the potential for increased exposure to weather-related (e.g., heat) extremes, producing both overlapping and interdependent risks for communities. Tornadoes on April 12th, 2020 in the US Southeast and extreme winds and rain on April 13th, 2020 in the US Northeast caused flooding and power outages, highlighting the urgency of examining ways in which cities are increasingly facing risks that overlap in space and time. Thus, strategies to build resilience to one event may actually decrease resilience to another event happening at the same time and place (Elmqvist et al. 2019). Interdependent vulnerabilities highlight the need to build adaptive capacity through a SETS lens and to address fundamental transformations in multiple social, ecological, and technological-infrastructure domains at the same time.

Recognizing the complexity of SETS dynamics provides a conceptual foundation for examining how SETS components can be mobilized together, and how they interact to generate resilience (Grimm et al. 2015; McPhearson et al. 2016; Grabowski et al. 2017; Markolf et al. 2018). In the URExSRN, the SETS framework explicitly acknowledges the interactions and interdependencies among the social–cultural–economic–governance systems (Social), climate–biophysical–ecological systems (Ecological), and technological-engineered infrastructural systems (Technological) that drive urban patterns and processes. Furthermore, these components need to be understood from a systems perspective to address key risks and understand where opportunities for solutions can be harnessed that synergize across S, E, and T dimensions of urban systems. Opening up the space to bring systems thinking into future visioning and scenario development can help us better understand the linkages, relationships, feedbacks, and planning in complex systems (Iwaniec et al. 2014). When we examine resilience strategies and visions from a SETS perspective, we can better see and imagine solutions to overlapping and cascading urban risks and vulnerabilities at multiple scales.

Clearly, the potential impacts of weather-related extreme events will add complexity to emergency response and require updated resilience planning. This means cities have to rethink adaptation and resilience strategies, such as using cooling centers to provide auxiliary cool space for those without air conditioning while also ensuring a safe space for aggregated residents during a pandemic. Increasing frequency and intensity of extreme events also highlights the need to move beyond emergency response toward building even longer term, anticipatory strategies to transform cities for disaster risk reduction. For example, co-produced scenarios in the URExSRN have imagined resilient urban futures where cities have invested heavily in green infrastructure as a nature-based solution to cool cities and reduce heat exposure, and decreased reliance on air conditioning and other twentieth century technologies for coping during extreme events. Transforming cities to be resilient to a potential extreme climate change future means thinking systemically and beyond status quo efforts to upgrade infrastructure or increase the availability of traditional technologies.

One of the key challenges cities face when designing, visioning, planning, or creating policy for resilient futures is the nature of uncertainty, surprise, and nonstationarity in the climate system, as well as the dynamic nature of complex urban systems (Box 12.1; Fig. 12.1). Traditional infrastructure aimed at mitigating the impacts of extreme events was designed according to predict-and-prevent logic rather

Fig. 12.1 Participants uncover randomly assigned "disaster cards" to consider features of resilience and how the scenarios withstand unexpected disturbances. Image credit: the authors

than a resilience-building logic, leaving cities vulnerable (Tyler and Moench 2012). Weather-related extreme events exist on the outer boundaries of meteorological distributions; the ranges of precipitation, temperature, and other weather phenomena that have been observed in the past (see Chap. 2). The dynamic nature of urban SETS now calls for more flexible and safe-to-fail designs (Ahern 2011) where both grey and nature-based infrastructure systems are conceived as flexible, multifunctional, and better to adapt to a world characterized by unpredictability. For example, a challenge facing urban infrastructure systems is that they are relatively inflexible, rigid, and long-lasting, including the institutions that manage and maintain them. Yet, there is uncertainty and changing levels of utilization, which occur during the lifetime of any infrastructure system (Chester and Allenby 2018). Typically, infrastructure systems are designed to meet demands decades into the future, but accurately anticipating, projecting, and capturing the complexity of future demands and system characteristics are tough—especially when disruptive technologies, climate uncertainty, and changing behaviors are considered (Markolf et al. 2018).

This complexity is not only true of infrastructure systems but also of governance systems. To build anticipatory capacity and resilience into governance and institutions across SETS will require very different ways of thinking about urban systems (Box 12.1; Fig. 12.1). We must also consider how to harness emergent properties of complex systems for resilience (Egerer et al., in press). In the URExSRN project, we bring the element of surprise and combined or cascading risks into scenario development precisely to ensure that visions articulated for resilient urban futures consider solutions that can be adaptive and flexible to uncertain futures.

Box 12.1 Using 'disaster cards' to identify features of resilience

The visions articulated by practitioner-researcher teams during scenario development workshops are subjected to large-scale disturbances or shocks to assess flexibility and the ability to adapt to uncertainty. During a visioning workshop in San Juan, Puerto Rico, a variety of disturbances ranging from energy, transportation, and communication disruptions, to increases in sea-level and temperatures, financial crisis, pandemic, mass migration to the mainland, and an influx of climate refugees were random-ly assigned as 'disaster cards.' The co-produced scenario would thus ex-perience large shocks in an unpredictable order, and workshop partici-pants were asked to assess how the social-ecological-technological sys-tems components depicted in the scenario fared. Asked to consider which parts are more or less affected and what mechanisms and attributes ac-tors conceded capacity to withstand, adapt, learn, or self-organize, the teams ultimately identified features of resilience before further refining the scenarios.

12.3 Future-Making as Privilege

If we understand our present conditions as the result of past decisions, the act of naming and imagining future possibilities needs to be recognized as a form of privilege. Future-making is an exercise in agenda setting; that is, the production of alternative futures sets the parameters of what is possible, even desirable, for a society. This is the purview of few. There is a need to reflect on the equity implications associated with the exercise of imagining positive futures as a way of influencing policy. Science fiction writer William Gibson said that the future was unevenly distributed, and, in the context of creating positive urban futures, the unevenness implies that not everyone has access to the tools and venues to imagine these alternatives. Indeed, we note that the ability to imagine alternative futures is offered to those who already enjoy a degree of influence in municipal decision-making and are therefore likely to reproduce the status quo (Turnhout et al. 2020; Jagannathan et al. 2020). It is imperative that marginalized communities are included in the process of imagining positive futures and have access, time, and resources to shape this conversation. The ability to think in the long-term is a form of privilege; those most vulnerable live in a reactive mode, needing to figure out how to survive the day to day, much less being able to plan for their future.

We cannot fully understand the nature of the problems, we are trying to solve for present and future generations unless we understand how problems—particularly climate inequity and environmental racism—were produced in the past. Through historical processes of economic change, ecological change, politics, and power, institutions have created vulnerabilities to weather extremes and inscribed climate inequities into the built environment (see Chap. 2). Since institutions are designed to replicate themselves, routinizing equitable solutions first requires examining the ways in which social separateness forms unequal climate burdens and is embedded in planning, policy, scientific, and economic practices. Therefore, future scenarios that are contextualized in a place-based historical awareness—in which people articulate not only the local narratives that form positive community identity and civic pride but also the exclusions and subordinations that persist today—hold more potential for shared resilient futures than do a historical scenarios.

Environmental justice theory offers conceptual tools for unpacking ways in which multiple and compounding forms of injustice exist in institutions (Bullard 1996). Future-making should link together material goals (such as energy affordability) with procedural goals (such as high rates of participation in the energy rate-making process among historically disenfranchised communities) that can sustain distributional outcomes (such as equitable access to low-cost renewable energy in all neighborhoods across a city). These linkages should be based on an understanding of ways in which distributional, procedural, and other forms of injustice are mutually reinforcing (Meerow et al. 2019). Vulnerability mapping and other tools for assessing environmental benefits and burdens (see Chap. 4) provide a starting point for articulating distributional patterns. These patterns of vulnerability express the ways in which structural inequality manifests in SETS. Visualizing these patterns enables

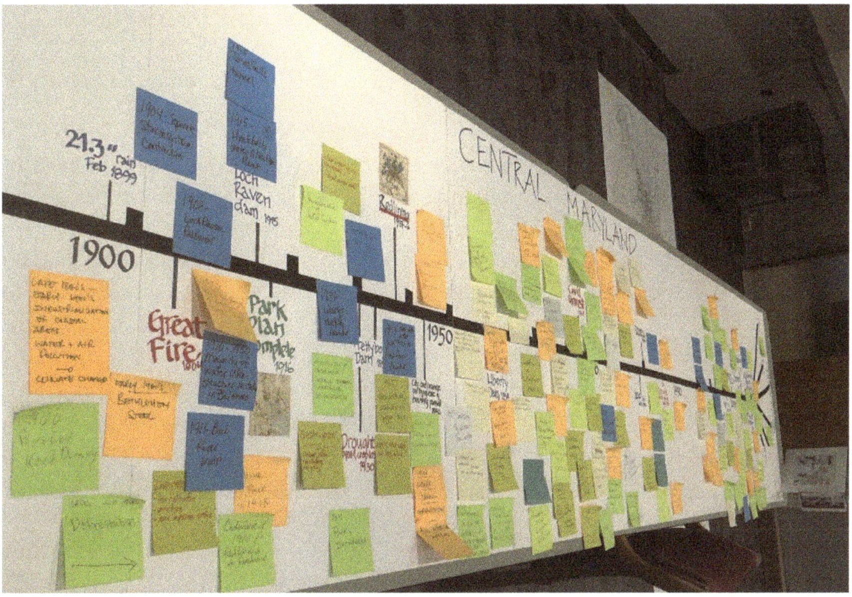

Fig. 12.2 Central Maryland timeline depicting multiple dam constructions, major drought (1930), and redlining from 1937 onward. Image credit: the authors

communities to consider how distributions may be misaligned with cultural values. The distributional patterns of vulnerability also provide an entry point for asking more fundamental questions about the processes that produce patterns of climate inequity and how to avoid reproducing such processes in the future (Box 12.2; Fig. 12.2).

Box 12.2 The importance of the past in navigating the future

To emphasize how the historical context of each city has shaped its current situation, URExSRN workshop participants reflect on major extreme events (natural and anthropogenic) extending back to the previous century. Building up to the present, these events are placed on a timeline and collectively examined to help identify correlations and any missing interdependencies between events. The notion of the future is thus conceptualized as an interpolation or 'linear unfolding' of the past. Across the network cities, past weather events and large-scale construction (namely dams and road infrastructure), alongside housing discrimination and residential segregation, have left low-income neighborhoods and communities of color disproportionately exposed to high heat, pollution, and severe weather. By jointly considering how past events, decisions, and actions have shaped the present, groups are better positioned to co-produce novel long-term goals for just and sustainable futures.

12.4 Developing an Urban Systems Science and Urban Systems Practice

The SETS framing brings forward a systems perspective that considers the reality of cities as complex systems and provides a baseline for developing a science of and practice for cities. A fundamental systems approach is essential in advancing our understanding of how to address critical challenges caused by the intersection of urbanization and climate change. It must be inclusive of social, ecological, and technological components, examine their interactions, and harness this complexity to imagine, plan, and develop strategies for more just and resilient futures. Given the urgency of issues, we collectively face to improve livability, justice, sustainability, and resilience in cities, bringing a systems approach to resilience planning and policymaking is critical, as is development of positive visions and scenarios that can provide more realistic and systemic solutions.

The complexity and diversity of cities inevitably bring up a plurality of perspectives, values, visions, and knowledge systems that define what cities are or should be (Muñoz-Erickson 2014). This places a demand for a transdisciplinary urban systems science and practice that harnesses and puts into action a diverse array of knowledge, rationalities, and ways of thinking to develop resilient futures. Scientific data, quantitative risk analysis, and computational models are important to such transdisciplinary science, but not sufficient. Intangible, nonmaterial flows and dynamics of urban systems, such as how different people experience risks or how they connect and interact with other groups in the city to build their social capital, are challenging to measure and model. James Scott (1998) uses the Greek term *métis* to describe the local practical knowledge that people, including those most impacted by changes in the environment, generate from learning and making sense of their contexts, but that cannot be codified or quantified. The diverse knowledge that city residents, business owners, planners, and professionals build from their lived experiences, including the practical strategies they employ locally to thrive in dynamic urban environments, is a crucial part of the *métis* needed to advance resilient urban futures. There is an urgent need for an urban systems science that includes these multiple forms of knowledge as legitimate and equal in the research process (Romero-Lankao et al. 2018).

The process of coproduction actively engages multiple voices and knowledge systems in the collaborative production of knowledge and solutions for urban sustainability and transformation. Collaborative approaches can be time consuming, politically uneven, and oftentimes messy (Turnhout et al. 2020). Yet, by integrating and explicitly deliberating diverse ways of knowing and perceiving, this approach makes the complexities, uncertainties, and needs of the system more evident than traditional planning and scientific approaches (see Chap. 7). This process also creates potential for a pluralistic urban systems science that engages decision-makers and local stakeholders directly in the creation of positive future visions, thus enabling an inclusive and anticipatory process to gain a more prominent role in urban practice and planning. The urban planning community, including decision-makers and local stakeholders, are fundamentally future-makers. Yet, in practice, the urban planning community is not making use of the wide array of anticipatory tools and approaches available to explore very long-term urban futures in the context of climate change (see Chap. 11). Understandably, planners and decision-makers prioritize risk-based assessments and near-term solutions in order to address urgent and pressing needs, and to do so within their terms of office. Although relatively rare, opportunities to craft long-term solutions of transformative change are essential for overcoming wicked, persistent, and emergent resilience challenges.

We seek to position the coproduction of resilient urban futures within urban systems science as an anticipatory knowledge practice to address the current deficit of futures thinking in urban planning and decision-making. Mainstreaming anticipation into organizational routine and practices requires institutional change. This may involve the redesign of a municipality's governance structure to harness and expand on already existing expertise and capacities across different units and departments, or the creation of new governance structures that explicitly embed future-making tools (i.e., scenario planning, storytelling, gaming, multicriteria assessments) and coproduction approaches into their organization's planning efforts (see Chap. 11). A key challenge is to go beyond one-time coproduction initiatives toward a routine practice within the organization's knowledge-making and decision-making processes such that strategies and solutions are not driven by crises; instead, they are continuously explored, stress-tested, and evaluated during times and spaces that allow for deep exploration of uncertainties and the generation of creative work.

At an individual level, engaging in the coproduction of urban futures allows actors to experience different roles and gain new capacities. Urban planners can gain core competencies of future-making, including creativity, imagination, and storytelling that enable visions to extend beyond extrapolation and into radical transformative pathways. However, although many planners have technical training in projecting future scenarios based on computational modeling, these models can still fall into the trap of representing future scenarios that simply conform to what has already been envisioned. Positive visioning processes are fundamental to comprehensive, neighborhood, and other types of planning, but if community leaders and residents are not engaged in imaginative work, then plans may lack ambition. In our work, community leaders and residents are civic knowledge producers, contributing crucial insights about how the city works and the vulnerabilities that its communities experience, and thus can help produce anticipatory knowledge alongside scientists and planning experts. Planners, scientists, community members, and residents alike can gain an expanded view of their cities through these processes and develop a future literacy that they can also take back to their respective communities. We suggest that futures work, which engages creativity, be explicitly embedded in professional practice, educational curriculums, and community capacity-building as a core competency to improve overall adaptive capacity.

To ensure broad representation, futures visioning work will need to creatively activate diverse technologies, designs, and engagement mediums. In the URExSRN, the scenario development process integrates creative storytelling through narratives, drawn vignettes, and design renderings, along with a data visualization platform for quantitative data, into the development of SETS future visioning (Fig. 12.3).

Fig. 12.3 Conceptual design illustrating aspects of a positive future vision for South Phoenix in 2080 captured during the scenarios workshop. *Image credit* Ian Klane

The increasingly wide availability of new technologies—including artificial intelligence, social media, gaming, creation of city digital twins, use of mobile apps to advance creative approaches, participatory mapping, and apps for sharing strategies and creating narratives and stories—provides new opportunities for engaging diverse knowledge systems and bringing positive visioning into workshops, community meetings, classrooms, and the private sector. These are also emerging as essential tools for engaging younger generations and bringing these important voices into coproduction of positive visions.

12.5 Positive Visioning for Resilience and Transformation

Overcoming persistent and emerging challenges, such as pervasive social and environmental injustice, require more than a responsive approach to change, instead demanding fundamental social, ecological, and technological transformations (Iwaniec et al. 2019). In turn, transformation requires major leaps forward and true game-changing strategies. Positive futures are critical to provide motivation, aspiration, and a basis to initiate real action and guide change. They serve as wayfinders to guide a course of action toward ambitious, positive trajectories that meet normative goals for society and the systems they are embedded in. Crafting visions of resilient urban futures through participatory processes fulfills a vital function in research, planning, and decision-making, providing a shared space to develop and assess strategies to transition from the current state to a desirable future state (Iwaniec et al. 2020).

Further developing an urban systems science that can provide actionable knowledge for decision-making is a transdisciplinary research agenda. It will require true boundary crossing to bring the knowledge, skills, tools, and ideas together in ways that can help achieve the normative goals and visions we have for our shared urban future. And yet for many city decision-makers, more discussion and focus needs to be placed on the value of scenarios, how to use them as a baseline for coproduction with communities, how to interpret them, and how to make sense of visions and projections that have embedded uncertainty. Not only do we need to advance an urban systems science for resilience, sustainability, and justice but also an urban systems practice, where decision-makers are thinking in systems, and where research is codesigned with practice.

Cities are already thinking about alternative and more desirable futures. Fundamentally, visions serve as a basis for all strategic planning in cities worldwide—covering scales all the way from the local, neighborhood level to city, state, and federal scales. Cities are being reimagined, reinvented, and shaped by dominant concepts and imaginaries that serve as a common vision to guide the visioning process and content of the visions. Can we utilize visioning processes as a driver of transformational change? Can we create an urban systems science that brings multiple forms of knowledge together? We argue that positive futures are critical to cocreating opportunities and generating realistic pathways for transformation toward sustainability. Coproduction enables incorporation of diverse sectoral, cultural, and disciplinary viewpoints into plausible and desirable future visions. Research and practice are beginning to create positive visions, develop future scenarios, generate pathways, create plans, and initiate implementation projects for improving urban sustainability, resilience, and human livelihoods in cities (McPhearson et al. 2017). This is encouraging but must be expanded. Positive futures are an opportunity to dig deeply into the key tensions and challenges to bring communities together to create shared visions or even to create pluralistic visions within which to reveal underlying conflicts, trade-offs, and tensions. Further, our approach and SETS framework provide key opportunities for building an urban systems science that can inform urban practice and, together, envision a positive, resilient urban future and chart pathways to get there.

References

Ahern J (2011) From fail-safe to safe-to-fail: sustainability and resilience in the new urban world. Landscape Urban Plann 100(4):341–343. https://doi.org/10.1016/j.landurbplan.2011.02.021

Bennett EM, Solan M, Biggs R et al (2016) Bright spots: seeds of a good anthropocene. Front Ecol Environ 14(8):441–448. https://doi.org/10.1002/fee.1309

Bullard RD (1996) Environmental justice: it's more than waste facility siting. Social Sci Q 77(3):493–499

Chester MV, Allenby B (2018) Toward adaptive infrastructure: flexibility and agility in a non-stationarity age. Sustain Resil Infrast 3(1):1–19. https://doi.org/10.1080/23789689.2017.1416846

Egerer M, Haase D, McPhearson T et al Urban change as an untapped opportunity for climate adaptation. Nat npj Urban Sustainability, *accepted for publication*

Elmqvist T, Andersson E, Frantzeskaki N et al (2019) Sustainability and resilience for urban transformations. Nat Sustain 2(4):267–273. https://doi.org/10.1038/s41893-019-0250-1

Grabowski ZJ, Matsler AM, Thiel C et al (2017) Infrastructures as socio-eco-technical systems: five considerations for interdisciplinary dialogue. J InfrastructSyst 23(4):02517002. https://doi.org/10.1061/(ASCE)IS.1943-555X.0000383

Grimm NB, Cook EM, Hale RL et al (2015) A broader framing of ecosystem services in cities: benefits and challenges of built, natural, or hybrid system function. In: Seto KC, SoleckiWD, Griffith CA (eds) The routledge handbook of urbanization and global environmental change. Routledge Handbooks Online, London and New York. https://doi.org/https://doi.org/10.4324/9781315849256.ch14

Iwaniec DM CDL, VanLehn K et al (2014) Studying, teaching and applying sustainability visions using systems modeling. Sustain Sci Pract Policy 6(7):4452–4469. https://doi.org/10.3390/su6074452

Iwaniec DM CEM, Barbosa O et al (2019) The framing of urban sustainability transformations. Sustain Sci Pract Policy 11(3):573. https://doi.org/10.3390/su11030573

Iwaniec DM CEM, Davidson MJ et al (2020) The co-production of sustainable future scenarios. Landscape Urban Plann 197:103744. https://doi.org/10.1016/j.landurbplan.2020.103744

Jagannathan K, Arnott JC et al (2020) Great expectations? Reconciling the aspiration, outcome, and possibility of co-production. CurrOpin Environ Sustain 42:22–29. https://doi.org/10.1016/j.cosust.2019.11.010

Markolf SA, Chester MV, Eisenberg DA et al (2018) Interdependent infrastructure as linked social, ecological, and technological systems (SETSs) to address lock-in and enhance resilience. Earth's Future 6(12):1638–1659. https://doi.org/10.1080/23789689.2017.1416846

McPhearson T, Pickett STA, Grimm N et al (2016) Advancing urban ecology toward a science of cities. BioSci 66(3):198–212. https://doi.org/10.1093/biosci/biw002

McPhearson T, Iwaniec D, Bai X (2017) Positives visions for guiding transformations toward desirable urban futures. CurrOpin Environ Sustainability 22:33–40. https://doi.org/10.1016/j.cosust.2017.04.004

Meerow S, Pajouhesh P, Miller TR (2019) Social equity in urban resilience planning. Local Environ 24(9):793–808. https://doi.org/10.1080/13549839.2019.1645103

Muñoz-Erickson TA (2014) Multiple pathways to sustainability in the city: the case of San Juan, Puerto Rico. EcolSoc 19(3). http://doi.org/https://doi.org/10.5751/ES-06457-190302

Romero-Lankao P, Bulkeley H, Pelling M et al (2018) Urban transformative potential in a changing climate. Nat Clim Change 8(9):754–756. https://doi.org/10.1038/s41558-018-0264-0

Scott JC (1998) Seeing like a state: How certain schemes to improve the human condition have failed. Yale University Press, New Haven, CT

Turnhout E, Metze T, Wyborn C et al (2020) The politics of co-production: participation, power, and transformation. CurrOpin Environ Sustain 42:15–21. https://doi.org/10.1016/j.cosust.2019.11.009

Tyler S, Moench M (2012) A framework for urban climate resilience. Clim Dev 4(4):311–326. https://doi.org/10.1080/17565529.2012.745389

Timon McPhearson is Director of the Urban Systems Lab and Associate Professor of Urban Ecology at The New School. He is a Research Fellow at The Cary Institute of Ecosystem Studies and Stockholm Resilience Centre and a member of the IPCC.

David M. Iwaniec is an Assistant Professor of Urban Sustainability at the Urban Studies Institute, Andrew Young School of Policy Studies at Georgia State University. He is a sustainability scientist researching anticipatory and systems approaches to advance urban sustainability, resilience, and justice. His work focuses on the co-development of scenarios and transition pathways for positive futures of urban transformation.

Zoé A. Hamstead is an Assistant Professor of Environmental Planning at the University at Buffalo and Director of the Community Resilience Lab. Her research builds on interdisciplinary approaches in urban planning, geography, and urban ecology to support climate equity and adaptive climate planning approaches. Dr. Hamstead teaches courses in environmental planning, environmental justice, economic concepts, geographic analysis of environmental problems, and engaged community energy studios. She holds a Ph.D. in Urban & Public Policy from The New School, a Master's in City & Regional Planning from the University of North Carolina at Chapel Hill, and a Bachelor of Liberal Arts from St. John's College.

Marta Berbés-Blázquez is an Assistant Professor at the School for the Future of Innovation at Arizona State University. Her research considers the human dimensions of social-ecological transformations in rural and urban ecosystems with an emphasis on vulnerable populations. Her work is informed by resilience thinking and political ecology at a conceptual level, and it is practically oriented toward qualitative, participatory, and anticipatory research methods. Specific topics of expertise include power dynamics and access in ecosystem services, scenario planning, resource extraction, eco-health, climate change adaptation, and transformation.

Elizabeth M. Cook is an Assistant Professor at Barnard College in the De-partment of Environmental Science. She is an urban ecosystem ecologist and her research focuses on future urban sustainability and human-environment feedbacks in urban and nearby native ecosystems. She con-ducts research on sustainability and resilience planning through participa-tory scenario development with local stakeholders. Her work seeks to un-derstand cities as social-ecological-technological systems with a compar-ative approach in Latin American and U.S. cities. Cook holds a Ph.D. from Arizona State University and a Bachelor of Arts from Wellesley College.

Tischa A. Muñoz-Erickson is a Research Social Scientist in the USDA Forest Service's International Institute of Tropical Forestry, in Río Piedras, Puerto Rico. She studies urban sustainability governance, including the policy networks, knowledge systems, anticipatory capacities, and strategies to advance sustainability, resilience, and equity. She is also actively involved in transdisciplinary platforms to facilitate the co-production of futures and transition pathways in the US and Latin American cities. Muñoz-Erickson leads the International Urban Field Station and the San Juan ULTRA and is also Co-PI of the National Science Foundation Urban Resili-ence to Extreme Events Sustainability Research Network (UREx SRN) and the SETS Convergence project.

Lelani M. Mannetti is a Postdoctoral Research Associate at the Urban Studies Institute, conducting comparative research across nine US and Latin American cities as part of the Urban Resilience to Extreme Events Sustainability Research Network. Alongside city stakeholders, she helps promote visionary thinking through the co-production of desirable and plausible future scenarios focused on resilience, sustainability, and equity. Dr. Mannetti is also a Fellow on the Intergovernmental Science-Policy Plat-form on Biodiversity and Ecosystem Services Values

Assessment, a methodological assessment of the diverse conceptualization of multiple values of nature and its benefits.

Nancy B. Grimm studies the interactions between climate variation and change, human activities, and ecosystems. Her interdisciplinary urban and stream ecosystem research focuses on disturbance, resilience, and bio-geochemical processes. Dr. Grimm is Regents Professor at Arizona State University and currently co-directs the Urban Resilience to Extremes Sus-tainability Research Network (UREx SRN). She was president (2005-06) and is a fellow of the Ecological Society of America, as well as a fellow of the American Geophysical Union (AGU), the American Association for the Advancement of Science (AAAS), and the Society for Freshwater Sci-ence. She is an elected member (2019) of the US National Academy of Sciences.

Correction to: Modeling Urban Futures: Data-Driven Scenarios of Climate Change and Vulnerability in Cities

L. Ortiz, A. Mustafa, B. Rosenzweig, Rocio Carrero, and Timon McPhearson

Correction to:
Chapter 9 in: Z. A. Hamstead et al. (eds.), *Resilient Urban Futures*, The Urban Book Series, https://doi.org/10.1007/978-3-030-63131-4_9

In the original version of the book, the chapter authors of Chapter 9 had missed to include their co-author Dr. Rocio Carrero to the author group unfortunately. The chapter has been updated now with Dr. Rocio Carrero's name and affiliation.

The updated version of this chapter can be found at
https://doi.org/10.1007/978-3-030-63131-4_9

Z. A. Hamstead et al. (eds.), *Resilient Urban Futures*, The Urban Book Series,
https://doi.org/10.1007/978-3-030-63131-4_13

Index

Z. A. Hamstead et al. (eds.), *Resilient Urban Futures*, The Urban Book Series,
https://doi.org/10.1007/978-3-030-63131-4